Zulu Plant Names

The old flower 'stalk' of *Gnidia kraussiana* looks like the tail of a leopard, hence its Zulu name **umsilawengwe** (leopard's tail) (drawing by Angela Beaumont).

Zulu Plant Names

Adrian Koopman

KZN PRESS UNIVERSITY OF KWAZULU-NATAL PRESS

Published in 2015 by University of KwaZulu-Natal Press
Private Bag X01
Scottsville, 3209
Pietermaritzburg
South Africa
Email: books@ukzn.ac.za
Website: www.ukznpress.co.za

ISBN: 978-1-86914-281-0

Managing editor: Sally Hines
Editor: Christopher Merrett
Proofreader: Catherine Munro
Typesetter: Patricia Comrie
Cover art: Angela Beaumont
Cover design: MDesign

Typeset in Times New Roman 10.5pt

Printed and bound in South Africa by Interpak Books, Pietermaritzburg

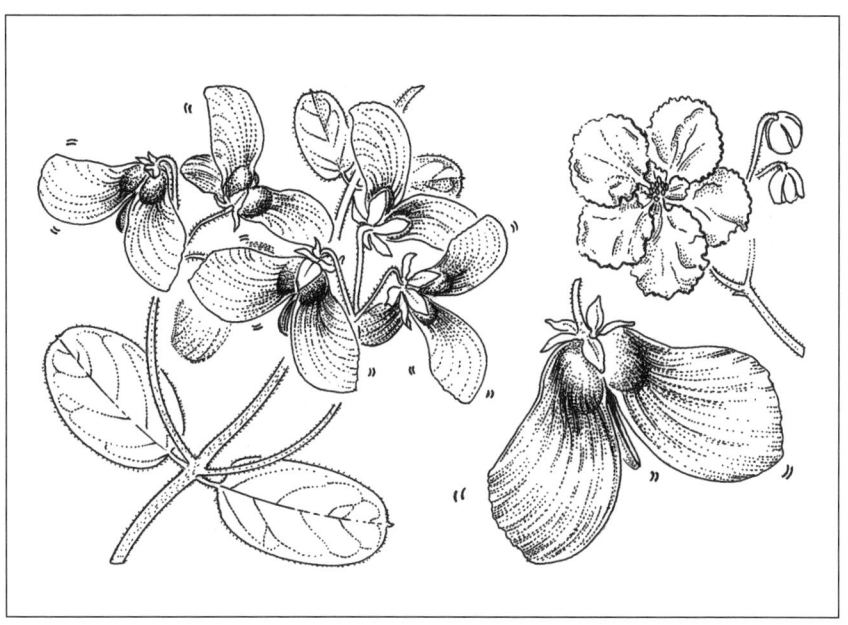

The winged fruits of the Lesser Moth-Fruit Creeper (*Sphedamnocarpus pruriens*), especially when in motion, give this plant the name **isinambuzane** (insect) (see page 132) (drawing by Angela Beaumont).

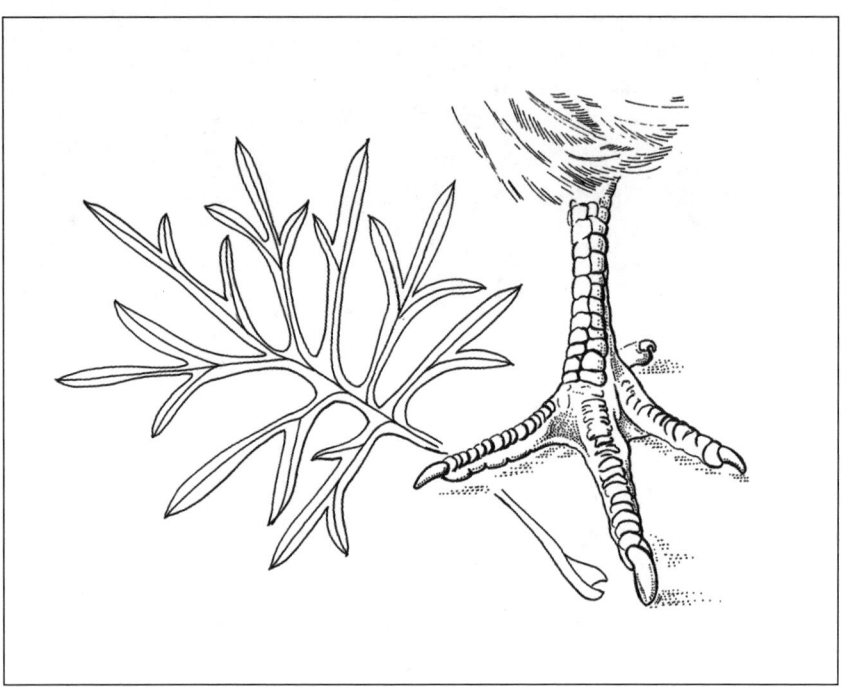

This drawing shows clearly why the Stalk-Flowered Pelargonium (*Pelargonium luridum*) has the Zulu name **unyawolwenkuku** (hen's foot) (see pages 77 and 137) (drawing by Angela Beaumont).

Contents

Explanatory preface

This is a book about the Zulu language. More specifically, it is a book about Zulu names for plants: trees, shrubs, herbs, grasses and so on. The emphasis throughout is on the relationship between the underlying meanings of the words that refer to plants, the plants themselves, and the ways in which they are both used and perceived in Zulu society. Many Zulu plant names do not have underlying meanings and thus little or nothing to say about the word-plant-culture interface. By and large these Zulu names have been left out of this book.

This is not a dictionary of Zulu plant names, nor is it a reference book of any kind. Although there is a strong focus on Zulu names, because this book deals with plants they are inevitably identified by their botanical names and, where relevant, names in English, Afrikaans, Sotho, Xhosa and so on. As much of my data came from collections of Zulu plant names that are 50 years old and older (often considerably older), many of the botanical names are out of date although every attempt has been made to provide the most recent version. In this book, a botanical name preceded by an asterisk indicates that the name given in the source has been superseded by a new name to be found in the list of synonyms after the bibliography. Thus an entry such as '**imfe**: Sweetreed (*Andropogon sorghum*)' indicates that *Andropogon sorghum* is no longer in current use. A check of *Andropogon sorghum* in the list of synonyms will show that this plant is now *Sorghum bicolor*.

At the time of writing (2011 to 2013) there was talk in the world of botany that some species of South African acacias would be reclassified in the genus *Vachellia*, so that *Acacia nilotica* would become *Vachellia nilotica*, *Acacia sieberiana* would become *Vachellia sieberiana*, and so on. As the proposed change was still being debated when this manuscript was nearing completion, it was decided to leave the several acacias in this book as they are. If the change goes ahead, the following species of *Acacia* would be affected: *A. ataxacantha*,

A. goeringi, A. karroo, A. luederitzii var. *retinens, A. natalitia, A. nilotica, A. robusta, A. sieberiana, A. swazica* and *A. xanthophloea.*

In the process of checking botanical names for both currency and accuracy, a number of errors have been discovered in the source books, particularly in Doke and Vilakazi's *Dictionary*. Rather than use the clumsy tag '[*sic*]' followed by a footnote giving the correct form, it was decided to leave such errors as given by the source, but put the correct version in square brackets after the erroneous form, with an equal sign, as in the following example:

> "species of forest tree *Protorhus longifolius*" [= *longifolia*] (Doke and Vilakazi 1958: 320).

Although all botanical names have been checked, there will inevitably be some errors. They remain the responsibility of the author.

Following standard convention, all botanical names are given in italics. As this book is primarily about Zulu plant names, they have been marked by the use of bold characters. All words in languages other than English (Afrikaans, Sotho, Xhosa, etc.) are in italics and this includes Zulu words if they are not the names of plants. Thus statements such as 'The name **umafuthasimba** (*Kedrostis foetidissima*) is derived from *amafutha* (fat) and *isimba* (excrement)' are common. The symbol < indicates 'derived from' while > means 'from which is derived' or 'gives rise to'.

The sources quoted in this book use a wide variety of styles regarding capitals, hyphens and spaces in the rendering of vernacular plant names, and no attempt has been made to standardise them.

Finally, not every Zulu plant name in this book has been assigned to a particular species. Occasionally Bryant or Doke and Vilakazi have given a plant name in their respective dictionaries, glossing it as "species of straggling bush" or "species of large forest tree". If the plant name is of sufficient interest and has an underlying meaning relevant to the themes of this book, then that name has been included even if the botanical identity of the plant cannot be pinned down. Examples are **ukhasikhulu** (big leaf), which Doke and Vilakazi (1958: 383) gloss as "sp. of large-leaved shrub"; **ilalamanzini** (what lies in water) for "a species of tree growing near water" (44); and **umchaphamanzi** (the water-lapper) for an unidentified species of fruit tree growing near water (109).

Acknowledgements

First and foremost I would like to thank Ndela kaBhekifa Ntshangase, a friend and a colleague for over 20 years in the discipline of isiZulu Studies on the Pietermaritzburg campus of the University of Natal, later the University of KwaZulu-Natal. I thank him for the years of wonderful discussions about the meanings of Zulu plant names, and the different ways in which plants are perceived and used in Zulu culture. He has shared freely his deep knowledge of Zulu culture and the natural world.

Then I would like to thank Noleen Turner (another isiZulu Studies colleague at the University of KwaZulu-Natal), Mkhipheni Ngwenya (botanist at the KwaZulu-Natal Herbarium, based at the Botanical Gardens in Durban) and Elsa Pooley (well-known botanist and botanical author). All three read earlier drafts of the manuscript and made pertinent and useful suggestions.

A great deal of thanks go to Angela Beaumont (botanist and botanical artist attached to the Botany Department on the Pietermaritzburg campus of the University of KwaZulu-Natal). I thank Dr Beaumont for her painstaking checking of every botanical name in the manuscript, and for providing the correct versions for plant names no longer in current use. I thank her also for her fine colour illustrations in this book, for the black-and-white pencil sketches and for her contribution of botanical terms to the glossary. Thanks are also due to Dr Benny Bytebier, the Curator of the Bews Herbarium (NU) of the University of KwaZulu-Natal and to Mrs Alison Young of the University of KwaZulu-Natal Botanical Garden for allowing Dr Beaumont to consult the botanical material in their collections for her illustrations

From the University of KwaZulu-Natal Press, I would like to thank Sally Hines (managing editor), Christopher Merrett (copy editor), Trish Comrie (typesetter) and Marise Bauer (cover designer) for their work in the editing and production of this book.

Finally, I would like to thank my wife, Jewel Koopman, for her help with the indexing, and for her encouragement and for her interest every time I brought her a new snippet of Zulu 'botanico-culture' to share.

The tree with no name

Hiram Wild in *A Southern Rhodesian Botanical Dictionary of Native and English Plant Names* writes about the tree **Schrebera mazoensis*, with the Shona name *muti-usinzita*:

> The Native name means "the tree without a name", and has a special significance in that it was the tree under which the old Baroswe crowned their Mambos or kings (1952: 122).

It is not clear whether "the tree without a name" was a name for a single specimen of this species, or for the species as a whole. Wild also gives the Shona name *mutisi* for this tree. Keith Coates Palgrave suggests that "the tree without a name" refers to one single specimen:

> One particular tree at Waddilove Mission in the Marandellas area in Rhodesia is called *muti-usina-zita*, meaning "tree without a name". The Rev. Brandon Graaf says that it is so named because it was thought not to occur elsewhere in that area and to have been introduced by the ancients who worked gold deposits there; it is also believed that it was through this specimen that God talked to his people in the days gone by. It should be noted that the name *muti-usina-zita* is also applied to *Cleistanthus schlechteri* but by the peoples in the Buhera district of Rhodesia (1977: 755).

While Coates Palgrave links the tree to the "ancients" and God talking to his people, and Wild links the tree to the crowning of kings, it is clear from both sources that this tree has a high symbolic and cultural importance among the community. In effect, while both naming and simultaneously not naming the tree, Shona culture is showing respect for it in the same way in which Zulu culture shows respect for certain culturally significant mountains by naming them *iNtabakayikhonjwa*: 'the mountain that is not pointed at', essentially meaning 'the mountain that is not named'.

Note that this showing of *ukuhlonipha* (respectful avoidance) towards the tree *Schrebera alata* is specific to the Shona culture. The Ndebele and Zulu cultures perceive this tree in different ways, with the Ndebele names *umbazankezo* (what carves wooden spoons) and *umnethamvula* (what makes the sky rain) and the Zulu name **umgwenyahlungulu** (**umgwenya** tree where the white-necked raven sits).

What is clear here in this story of the "tree without a name" is the relationship between plant, language and culture. The examination of this relationship is the main theme of this book on Zulu plant names.

1 Introduction

Background to this book

In 2002 the University of Natal Press[1] published my book *Zulu Names*. As the title implies, this book is about proper names in the Zulu language. A section on anthroponymy (personal names) covers topics such as given first names, 'colonial' names, surnames and clan names, clan praise names, and the names of groups such as regiments. Another section covers toponymy (place names) from various aspects. Yet another section covers a variety of names, including the Zulu names of schools, shops and homesteads. In this section there is also a chapter on the Zulu names of birds. In the original manuscript submitted to the University of Natal Press for consideration there was a chapter on the Zulu names of plants and at about 40 pages it was the longest chapter in the book. For various reasons, it was felt that plant names did not belong in that book and it was suggested that I resubmit it at a later stage as a possible separate publication.

In order to do so, the material originally collected on Zulu plant names would have to be put into context. My original chapter had brief sections on the morphology (structure) of plant names as well as some discussion on the different layers of meaning found in plant names. This book has separate chapters on structure and meaning and these chapters introduce the linguistics disciplines of morphology and semantics as they apply to both the Zulu language and to plant names in that language.

Since collecting my earlier data in the late 1990s and writing it up as a chapter for the planned Zulu name book, a great deal of research has been

---

1. Now the University of KwaZulu-Natal Press.

done into the nature of indigenous knowledge systems (IKS) in South Africa. In KwaZulu-Natal much research has also been undertaken into community-held knowledge of plant names, especially under the auspices of the Zulu Botanical Knowledge Project (ZBKP). Both of these, IKS and the ZBKP, are researched in some detail in Chapter 2 of this book. IKS and folk-taxonomy of plants are compared and contrasted with Euro-Western knowledge systems, and the system of plant taxonomy known as the Linnaean system. The history of botanical studies in the Western world is also examined, with emphasis on the ways earlier scholars dealt with the problems of correlating names and plants.

Anna Pavord's excellent book (2005) on the history of botanical studies makes it clear that much of the earlier study and classification of plants was done by apothecaries, who needed to know more about plants for their medicinal use. Even today in KwaZulu-Natal by far the bulk of community-held plant knowledge lies with Zulu traditional doctors or herbalists (*izinyanga*) who play the same role vis-à-vis plants and healing as the apothecaries of medieval Europe. For this reason, or rather for these reasons, I have included a chapter in this book on the use of 'muti' (traditional medicine) in Zulu culture.

Most of this book contains the bulk of my earlier research into Zulu plant names, considerably expanded and divided into chapters according to both the underlying meaning of plant names and the ways in which plants are used in Zulu society. The longest chapter contains the names of plants that talk about the plant itself (its appearance, habitat and so on). Then there are chapters that deal with plants used for medicinal purposes, as protective and love charms, and for practical purposes.

In the final chapter I look at both the power of plant names, especially in ritual chanting, and the use of plant names as mnemonics in a traditional oral society.

When is a name not a 'name'?

In *Zulu Names*, the chapter on bird names is the only one where I do not talk about 'proper names' (or 'proper nouns'). The term 'proper noun' goes back to traditional English school grammar of the first half of the twentieth century. 'Proper nouns' refer to nouns usually spelt with a capital letter in English such as Mary, Churchill, Peter Pan, London, Church Street, Kilimanjaro, the Thames, Pepsodent and Ryvita. Traditionally, proper names as they are sometimes called refer to the names of people (both real and fictional), the names of places and geographical features, brand names and a few other categories. There is often debate about whether or not the names of the days of the week, the months of the year, and the four seasons are proper names.

Proper nouns are contrasted with 'common nouns', nouns that are <u>not</u> proper names, such as 'aardvark', 'bacon', 'democracy', 'butterfly', 'sleep' and 'daisy'. As we see in this list, 'aardvark', 'butterfly' and 'daisy' are common nouns and not proper names. They can of course be <u>used</u> as proper names. 'Aardvark' could be given as a nickname to a clumsy, lumbering type of person who is only active at night; 'Daisy' is a relatively common English name for a girl, as indeed are 'Rose', 'Iris' and 'Lily'.

Words for birds such as 'kingfisher', 'hawk', 'robin', 'eagle', 'wren' and 'sparrow' are common nouns, not proper names. Even if spelt with capital letters such as Namaqua Sandgrouse, Crowned Plover, House Crow and Redbilled Hornbill,[2] these are still not proper names. Zulu words like *umkholwane* (hornbill), *iphothwe* (bulbul), *intengu* (drongo) and *intshe* (ostrich) are likewise common nouns and not proper names. And yet MacLean under the heading 'names' makes statements such as "The scientific names of birds in this book . . .", "English names are mostly those . . .", "the Afrikaans names . . ." and "Bird names in the African languages . . ." (1985: xxiv). Most people would surely be perfectly happy with the use of 'name' and 'names' in this manner. But from the point of view of the onomastics scholar,[3] we need to be aware that there is a difference between 'parrot' and 'Polly' in the sentence 'This is my pet parrot Polly'. 'Polly' is a proper name, 'parrot' is not.

So to come back to the statement that I made above about the chapter on bird names being the only one not about 'proper names', it is indeed the only chapter in a book about Zulu proper names that concerns Zulu common nouns. And the chapter, ostensibly about Zulu plant names, that never made it into *Zulu Names* is in fact about Zulu common nouns referring to plants. Humans frequently give proper names to animals, especially domestic animals and animals kept as pets. Even wild animals are given proper names if for some reason or other they become identifiable individuals that interact with humans in some way. Birds are occasionally named, again especially if they are kept as pets or have a close domestic relationship with humans. But I am yet to come across any case of a proper name given to a plant by a human. 'This is

---

2. Examples from MacLean (1985).
3. Onomastics is the accepted academic term for the study of proper names and naming systems.

my pet parrot Polly' is a statement that does not raise any eyebrows. But the following would certainly do so:

> 'That's a splendid plant you have growing there. What's it called?'
> 'Oh, that's our favourite aspidistra. We call her Annie.'

As in the case of Gordon MacLean's book on southern African birds, so botanical reference books also refer to the 'names' of plants. Richard Boon (2010: 18–19), for example, talks of both the 'scientific names' and the 'common names' of the trees of eastern South Africa.

Botany and linguistics

Let us consider the following, all referring to the same species of tree:

- *Clutia abyssinica*
- Large Lightning Bush
- *grootbliksembos*
- **umembesa**

Whether we call these 'names', 'nouns' or just 'words', they are all items of language. We have four distinct words or phrases here: *Clutia abyssinica* is the Latin-based botanical or scientific name of a particular plant. It is a binomial or 'two-part name', usually printed in italics. The first half (the genus name) is always given with a capital letter, while the second half (the species name) is always given in lower case. All botanical names in this book are printed in italics. Large Lightning Bush is a phrasal name; that is, it consists of more than one word. It is the English common name for the same plant and this book follows Boon (2010) in using standard font with upper case. *Grootbliksembos* is the Afrikaans common name. It is also a phrasal name in that it consists of *groot* (large) + *bliksem* (lightning) + *bos* (bush), but its phrasal nature is obscured by the tendency of the Afrikaans language to consolidate such phrases into single and comparatively long compound nouns. **Umembesa** is the Zulu common name. In this book I use bold font for Zulu plant names. Names in Xhosa, Southern Sotho, Swati and Tembe-Thonga will be given in italic font with the abbreviations Xh., SS, Sw. and TT.

All nouns, in any language, have three 'aspects'. First, they have a form, a shape, a structure. In their spoken form they consist of a string of phonemes, or speech sounds, and these may be grouped together into syllables. In their written form, they consist of a string of letters. The strings of spoken sounds and the strings of written letters both constitute words, spoken and written. These words can themselves be broken down into smaller parts, as we have already seen with *grootbliksembos*. The Zulu **umembesa** consists of the noun class prefix **u-** and the stem **membesa**. Words may also be grouped together to form phrases, such as 'Large Lightning Bush'.

Second, all nouns have a 'referent'. This is the 'thing', 'state', 'condition' or 'phenomenon' in the real world referred to by the noun. It is that which is denoted by the noun. The four names given above have distinctly different structures, but they all have the same referent, which according to Boon (2010: 232) is a deciduous shrub or tree found in forest understories and margins from the Eastern Cape to Mozambique and Zimbabwe, and generally widespread in Africa. Further details about its stem, leaves, seeds and fruit, together with photographs of this tree, help to identify it as a species.

Third, all nouns have meaning. Indeed, as we shall see in Chapter 3 on semantics and in the second half of this book, nouns may have many levels and many types of meaning. The one that concerns us here is the 'referential meaning' or 'denotative meaning'; in other words the meaning that allows the person using the plant name to associate it with a particular species of plant. To illustrate, the Zulu noun **umembesa** has the referential meaning 'the tree species known to botanists as *Clutia abyssinica*, or a single example of this species'.

This book is about names, about words that are linguistic items. But as almost all the names discussed in this book have one thing in common, and that is that their referent is a plant, a botanical item, it is also willy-nilly a book about botanical issues. You could say that this book is about the interface between linguistics and botany – how words and plants relate to each other. It is also, of course, a book about humans, for words are created and used by humans and they only give names to plants they know, identify and use. We must also remember that in order 'to make sense out of chaos',[4] the relationships between plants and other plants, between words and other words, and between plants and words are created and described by humans. Categories (whether botanical or grammatical), hierarchies and taxonomies, the concept of genus and species, and decisions on what to put into dictionaries and what to leave out are all human constructs.

---

4. A phrase and concept that will appear a number of times in Chapter 2 when the history of botanical taxonymy is discussed.

One of the problems about a book on the interface of linguistics and botany is that there are a number of 'technical' terms shared by both disciplines, but with variant meanings. For example, earlier I said, 'The Zulu **umembesa** consists of the noun class prefix **u-** and the stem **membesa**.' In linguistics generally, and in grammatical descriptions of Zulu nouns specifically, a 'stem' is what is left of a noun when its noun class prefix is removed. In the words *umuntu* (person) and *abantu* (people), *-ntu* is the stem with the meaning 'human', and *umu-* and *aba-* are prefixes with the meanings 'singular' and 'plural' respectively. 'Stem' is also a botanical term. When Boon describes the stem of the Spiny Splinter-Bean (*Adenopodia spicata*) as "light-brown, deeply fissured" (2010: 190), he is using the word 'stem' in a different way to the one I used in 'the noun class prefix **u-** and the stem **membesa**'.

For purely linguistic reasons, Zulu grammars describe nouns (and other parts of speech) as having 'stems', whereas verbs are described as having 'roots'. A Zulu verb used in a sentence, as opposed to its listing in a dictionary, can be a very complex structure. The Zulu word *akasangibhaleli* is often used to describe the agglutinative structure of Zulu words.[5] It translates into English as 'he no longer writes to me', and the structure of the word is [*a* (not) + *ka* ([not] he) + *sa* (any more) + *ngi* (me) + *BHAL* (write) + *el* (to) + *i* (not)]. In this conglomerate of morphemes, the ones in lower case are grammatical affixes, the one in upper case is the verb <u>root</u>.[6] In Zulu grammar, roots can variously be 'monosyllabic', 'vowel-commencing', 'derived from isiXhosa' and so on. Elsa Pooley (1998) describes roots (*inter alia*) as "fleshy, clustered", "fibrous, rooting from nodes", "tuberous" and "woody". Clearly in this book we have two different kinds of root.

Two other terms that overlap are 'compound' and 'morphology'. 'Compound' basically means 'having more than one identifiable part in a single structure' and this can apply equally to linguistics (compound nouns, verbs, adjectives) and botany (compound flowers, leaves, fruits and seeds). The word 'morphology' likewise has a single meaning with dual application. It simply means 'structure' or 'study of structure'. In linguistics, 'morphology' usually refers to the structure of words rather than smaller structures such as phonemes or larger structures such as sentences. Morphology in the botanical

---

5. Agglutinative describes a language that combines a number of grammatical items in one word, as in Zulu *akasangibhaleli*. English, which uses six separate words to say the same thing, is not an agglutinative language.
6. A morpheme is a word part with a separate meaning, whether grammatical ('not', 'past', 'conditional' and so on) or lexical ('real life').

sense refers to the larger elements of a plant: habit (that is shape and size of the whole plant), leaves, stems, inflorescences, bark, roots and flowers.

Sources of data

Although I have spent quite some time in the field recording Zulu place and bird names from knowledgeable Zulu-speaking sources, my recording of plant name data from oral sources has been minimal. My university colleague Ndela Ntshangase, who has considerable knowledge of the names of plants used for 'muti' purposes, has kindly shared much of his knowledge with me. But the bulk of the information about Zulu plant names in this book has come from written sources, mainly Zulu dictionaries and works of a botanical or botanico-medical nature that have extensive lists of Zulu plant names. Let us look at these, in chronological order.

Father A.T. Bryant's *Zulu-English Dictionary* contains a number of entries for Zulu plant names. Although his list of names is by no means as comprehensive as later works, Bryant's dictionary contains much socio-cultural material relating to plants rarely found in other publications. Consider, for instance, the entry for **umunga**, a name given for a number of acacia species:

> **umu-Nga** . . . Several kinds of mimosa thorn-bushes (*Acacia horrida* or Doornboom; *A. Natalitia*;[7] *Dicrostachys nutans*, etc.)
> Phr. *qibugele! ng'elomunga! ng'elomtolo! lidhla ligodile intanyana!* hurrah! it's an *umunga* (that is one thriving on the *umu-Nga* bush)! it's an *umtolo* (or one thriving on the *umTolo* bush)! it eats with its little neck hidden away! – cry of a boy who has found an *i-cimbi* caterpillar, which are eaten with great delight.
> *N.B.* The *umu-Nga* bushes, like the *euphorbia, um-Kiwane*, etc. have the reputation of being dangerously attractive to lightning, which apparent superstition, from the resinous nature of their wood, may probably be a truth (1905: 416).

Bryant's 'Zulu Medicine and Medicine Men' (1909) was republished in book form by Struik in 1970 and its dust cover says:

---

7. Early writers, including Bryant (1905) and Doke and Vilakazi (1958), commonly wrote the specific epithet with an initial capital if it was based on a proper name, as in *Natal* in this case. The convention today is to write all specific epithets with a lower case initial letter, irrespective of the origin.

During his stay of 45 years amongst the Zulus Dr. Bryant kept accurate notes of how the Medicine-man was trained, his healing methods and the medicines he used. He made a special study of the healing power the Zulus attributed to certain parts of trees, shrubs and plants . . . A table of Zulu Medicinal plants, with the Zulu names, Botanical names and their Use, as mentioned in Dr. Bryant's Ms., has been added to this book.

The earliest list of Zulu plant names to appear in a botanical work[8] is in Bews' *An Introduction to the Flora of Natal and Zululand*. His section headed 'List of Zulu Plant Names' (1921: 455–467) contains some 850 Zulu plant names. In a footnote Bews says he was greatly assisted by Bryant's dictionary (1905) and also by Bryant's 'Zulu Medicine and Medicine Men' (1909). Bews also notes he "had the opportunity of consulting a manuscript list compiled by Mr. J.S. Henkel" that I have not been able to trace.

John William Bews (1884–1938) was professor of geology and botany at Natal University College in Pietermaritzburg and its first principal (Guest 2010: 3). He is commemorated in the name of the John Bews life sciences building on the Pietermaritzburg campus of the University of KwaZulu-Natal, in the plant genus name *Bewsia*, and in the two species names *Gymnostomum bewsii* and *Lophocolea bewsii*.

Father Jacob Gerstner (1888–1948) was a Bavarian botanist and Roman Catholic missionary who emigrated to Africa in 1924. He published 'A Preliminary Check List of Zulu Names of Plants' in several volumes of the journal *Bantu Studies* from 1938 to 1941. Gerstner's contributions to botany are recognised in the plant names *Aloe gerstneri* and *Pavetta gerstneri*. It has proved difficult to obtain details of Gerstner's life as a missionary-botanist in South Africa, but here is a brief mention from Boon, curiously enough under the entry for the Maputaland Croton (*Croton steenkampianus*) named after

Dr Steenkamp MP for Vryheid who helped Rev Abbé Gerstner. Gerstner (1888–1949)[9] was one of the first systematic gatherers of ethnobotanical information in Zululand (2010: 236).

---

8. That I have come across, that is! Blessing's list of Zulu medicinal plants, collected between 1901 and 1904, was published only in 2012.
9. Gerstner's correct dates should be 1888–1948.

Gerstner himself in the preface of 'A Preliminary Check List' thanks Professor Mogg of Pretoria, a Miss Dr Weintroub of Johannesburg, Mr Boocock (the Conservator of Forests in Natal) and the Reverend Willibrord Binder of Centocow Mission (Natal) for "allowing him to use their lists of Native names of plants" (1938: 216). It is clear that at this time many people were involved in the collection of vernacular names of plants.

Gerstner's list is particularly useful because in many cases he records the district where plant names were collected, giving an indication of regional variations. Amusingly, for abbreviations of the districts of Natal and Zululand he uses motor car number plates. These are still in use in KwaZulu-Natal today and to any present resident this kind of shorthand is very familiar: NE for Estcourt, ND for Durban, NP for Pietermaritzburg, and so on. Gerstner also often includes snippets of ethnobotanical information. The following are typical entries:

isAgude (4 NZ, NUF),[10] *Strelitzia augusta Thb., usually called isiGceba. They use the leaves to make strings for building native huts (1938: 218).

umBangazi (fairly general in Zululand with the exception of Wome-bush where I found umVangazi). *Albizia mossambicensis Sim. As the meaning of this name is more or less the same as umBangabanga, you may find this name used for other quickly-growing trees like Macaranga capensis, *Trema bracteolata and *Ekebergia velutina as well (1938: 221).

The Southern-Sotho-English Dictionary by Adolphe Mabille and H. Dieterlen has been useful for examples of Sotho plant names. In the Introduction we learn:

Mme H. Dieterlen . . . because of her great desire to help her husband in this very important work of his, began collecting plants with their Sotho names. Then she had them identified and contributed several hundreds of plant names, an invaluable piece of work (1950: v).

Two years later, Hiram Wild produced A Southern Rhodesian Botanical Dictionary of Native and English Plant Names. This has been useful in

---

10. uMthunzini and uMfolozi districts.

providing names in Zimbabwean Ndebele and the various dialects of Shona. Wild refers to "numerous collectors [who] have over many years collected the native names of plants and these have been filed in the Government Herbarium" (1952: i). His list of the names of such collectors is extensive and it is clear that in the first half of the twentieth century there was a great deal of fieldwork being undertaken to collect vernacular plant names in what was then Northern and Southern Rhodesia (now Zambia and Zimbabwe). Wild was not quite as lucky as his contemporary, the Reverend Dieterlen, whose wife personally collected several hundred plant names specifically for his dictionary. Nevertheless, Wild's wife did at least "cheerfully [take] over the tedious task of typing the manuscript" (1952: ii).

In my 45 years or so as a student of the Zulu language, no other reference work has been as important to me as the *Zulu-English Dictionary* by Clement Doke and Benedict Vilakazi (1958).[11] The acknowledgements in the Introduction take over a page (xii–xiii) and among these:

> Special mention must be made of the help of the Rev. J. Gerstner, Ph.D., in placing at our disposal his published and unpublished collections of botanical terms, a large number of which have been incorporated in the dictionary. Where he himself had insufficient verification and where we could get no further confirmation, botanical entries were discarded. In addition Gerstner supplied considerable lists of natural history terms.

Eve Palmer and Norah Pitman brought out *Trees of Southern Africa* in 1972. In three volumes, this monumental work has extremely useful material on the various ways people utilise trees, including both practical and cultural uses. It is also unique for the amount of detail given to the relationship between trees and animal life, including insects, birds and mammals. Palmer and Pitman provide names in various Bantu languages, so all the trees occurring in KwaZulu-Natal have Zulu names, usually more than one.

Keith Coates Palgrave's *Trees of Southern Africa* (1977) is likewise rich in ethnobotanical and cultural information, but includes no vernacular names, which makes this well-known work of limited use for a book on Zulu plant names. I have used Coates Palgrave to corroborate or further illustrate the

---

11. This dictionary was originally published in 1948 and revised in 1953. My copy is a 1958 reprint of the revised edition.

work of other authors in the second half of this book when writing about the medicinal, 'magical' or practical use of certain plants. A new edition of this work, revised and updated by Meg Coates Palgrave, was brought out in 2002.

Eugene Moll's *Trees of Natal* (1981) gives Zulu names for almost every tree mentioned in the book. Often two, three or even four Zulu names are recorded for a single species. It is a pity Moll gives no indication where these Zulu names originated. His acknowledgements make no mention of Zulu names and the section headed 'Zulu Names and how to use the Zulu Index' (1981: xxv) gives no indication of sources. In contrast to Coates Palgrave, Moll provides no ethnobotanical or cultural information, so I have not made a great deal of use of his data, apart from the occasional cross-reference.

Ethnobotanist Anthony (Tony) Cunningham completed his doctoral research into the use of indigenous plant material on the Maputaland coastal plain in 1985. The research concentrated on the tapping of palms for wine, the use of palm leaves, grasses and rushes for weaving, the use of trees for firewood, construction and carving, and so on. He gives both Zulu and Thonga names for plants employed in this way.

The year 1993 marks the start of the Elsa Pooley era, with the publication of *The Complete Field Guide to Trees of Natal, Zululand and Transkei*. In this book Pooley combines an extensive listing of Zulu names for trees with ethnobotanical and cultural information, thus combining (for my purposes at least!) the best of her predecessors Coates Palgrave and Moll. In her acknowledgements Pooley thanks thirteen individuals "for lists of Zulu, Xhosa and Thonga names" (1993: 8).[12]

Pooley's book on trees was followed by *A Field Guide to Wild Flowers: KwaZulu-Natal and the Eastern Region* (1998). As with the earlier work, this offers the student of Zulu plant names and Zulu ethnobotanical practices a wealth of information. Plant names for each species are given (where available) not only in Latin, English and Afrikaans, but in Zulu, Xhosa, Southern Sotho, Swazi and Thonga. In her acknowledgements Pooley refers to "Roddy Ward [who] shared his work on Zulu names of plants". She also mentions that "C.J. Skead's list of Xhosa plant names and uses was incorporated" and that "Gideon Dlamini, Department of Agriculture, Swaziland, supplied a list of Swazi names" (1998: 11).

---

12. My copy is the third impression, dated 1997.

The third book of this enormously productive Pooley era was *Mountain Flowers: A Field Guide to the Flora of the Drakensberg and Lesotho* (2003). While this work contains essentially the same useful numbers of vernacular plant names and ethnobotanical information, it has been of less value to me than the two earlier works, simply because in the geographical region covered, Southern Sotho is by far the majority African language. If I were to write a book on Southern Sotho plant names, I am sure this contribution of Pooley's would become as well-thumbed as the earlier two volumes.

Anne Hutchings' *Zulu Medicinal Plants* (1996) has been a particularly useful source, ranking with Doke and Vilakazi and the Pooley books as the most productive sources for this book on Zulu plant names. The value in Hutchings is several-fold. She includes all botanical types – trees, shrubs, herbs and all other types of plants – and socio-cultural as well as medicinal uses, often making reference to other cultural groups such as the baSotho. Her description of medicinal uses is astonishingly detailed and contains substantial chemical notes; and, above all, her lists of names (botanical and English, Afrikaans and Zulu vernacular) are far more extensive than those of any other source. Occasionally she includes several forms of the same Zulu name where one is accurate and the others are misspellings or otherwise incorrect, but this is no problem because it is usually very easy to see which is the correct Zulu form. I also appreciate that Hutchings regularly gives earlier synonyms for botanical names, a feature not often found in books with a botanical focus.

Boon's *Pooley's Trees of Eastern South Africa* (2010) is, as the name indicates, a new edition of Pooley's earlier 1993 work on the trees of the erstwhile Natal, Zululand and Transkei.

In 2011, the mammoth *Dictionary of Names for Southern African Trees* was published, authored by Braam van Wyk, Erika van den Berg, Meg Coates Palgrave and Marie Jordaan. It gives tree names from no less than 27 southern-African-origin languages as well as Afrikaans, German, English and Latin. It has been of minor use to me as a source of Bantu cognates for the Zulu tree names.

The year 2012 saw the publication of *Medicinal and Charm Plants of Pondoland* by Sinegugu Zukulu, Tony Dold, Tony Abbott and Domitilla Raimondo. Sixty plants found in the Pondoland region have been selected and each has been given a page, with a colour photograph (or photographs) accompanying explanatory text that covers briefly the appearance of the plant, its medicinal use or use as a charm, and in many cases an explanation of the underlying meaning of the Pondo name. In some instances, these underlying meanings are linked to the use of the plant, for example the name *iphamba* (*Polystachya pubescens*) for which the authors explain:

> It is believed that sorcerers are able to control and direct lightning towards their victims and *iphamba* is used to 'confuse' or divert lightning strikes away from its intended victims. *Iphamba* derives from *ukuphamba*, meaning 'to dodge or outwit' (2012: 32).

In other instances the name is interpreted, but without any such links, as in the name *umgadankawu* (*Albizia adianthifolia*): "*Umgadankawu* derives from *ukugada*, meaning 'to guard' and *inkawu* meaning 'monkey'" (2012: 23); and in the name *umayibophe wehlathi* (*Acridocarpus natalitius*): "The name derives from *ukubopha*, meaning 'to knot or tie up', and *ihlathi* means 'forest'" (2012: 25).

The Pondo language is very similar to the Zulu language and Pondoland lies to the immediate south of the Zulu-speaking region, so there is a considerable overlap of plant species, of cultural beliefs and plant uses, and in names and their underlying meanings.

Also published in 2012 was a book on Zulu plant names by Norwegian doctor Henrik Greve Blessing. Description of Blessing's work is a fitting way to round off this section on contributions to the collection of plant names in indigenous African languages as although the book appeared in 2012, the research was done well over 100 years before. Authored by B.S. Paulsen, H. Ekeli, Q. Johnson and K.R. Norum, the book tells of the visit to what is now KwaZulu-Natal by Blessing between 1901 and 1904. Unable to find medical work, Blessing instead contacted local Zulu healers and learned from them about the plants they used for medicinal purposes. What makes this publication particularly interesting is that it appears in two volumes. One is a facsimile reprint of Blessing's fieldwork notebook, with notes (mostly in Norwegian) in his original handwriting; the other is a transcription and translation into English of his notes. To this the authors have added extensive photographic illustrations of the 97 plants in Blessing's notebook, as well as details of recent medical research into these plants, of which the following is an example:

> The hexanic extract of leaves from V[ernonia] colorata increased blood glucose in normoglycaemic rats in a dose-dependent way, while acetone extracts prevented glibenclamide induced hypoglycaemia, suggesting that the extracts have the opposite effect on basal blood glucose in normoglycaemic rats (2012: 92).

Blessing's list of plant names contains nothing new, in the sense that all the Zulu names he gives are recorded in Bryant, Gerstner, Pooley, Hutchings and the other authors mentioned above who have listed Zulu plant names. What makes the Blessing publication useful is the inclusion of the original field notes. The way in which these have been transcribed shows clearly the problems involved when oral information is transferred to field notes and these are later published.[13]

Cross-language borrowing (the 'Tambootie Factor')

It is generally accepted that when two or more languages share the same space, or are spoken in contiguous areas, there will be a great deal of mutual influence. One of the most obvious ways in which languages influence each other is in the 'borrowing' of vocabulary. These are sometimes known as 'loanwords' or 'adoptives'. To give a few examples, English has adopted the following words from Zulu: *impala*, *inyala*, *mamba* and *donga*. Even the slang word 'babbelaas', meaning a hangover from excessive drinking of alcohol, is an adopted version of the Zulu word *ibhabhalazi* with exactly that meaning. The world of plant naming is by no means free of this phenomenon and under this heading we shall look at a variety of Zulu plant names adopted from other languages (mainly English and Afrikaans) and also at a number of English and Afrikaans vernacular names, as well as botanical names, derived from Zulu names.

Zulu names adopted from English or Afrikaans

It is to be expected that where a plant has been introduced as a newcomer to Zulu-speaking areas there will be no Zulu name for it and a 'Zulu-ised' version of an English or Afrikaans name will be used. When one looks at trees, vines and bushes producing fruit new to the Zululand-Natal area in the early 1800s, one sees many names such as **i-aphula** (< apple), **iwolintshi** (< orange), **umalibele** (< mulberry), **ugwava** (< guava), **u-anyanisi** (< onions), **utamatisi** (< Afrik. *tamatie* tomatoes) and **ugilebhisi** (< grapes). In the same way the Seringa (*Melia azedarach*) is known in Zulu as **umsilinga**. Note that in words like *iwolintshi*, *umalibele*, *ugilebhisi* and *umsilinga*, the phoneme /r/ in the English word has been replaced by /l/ in Zulu.

---

13. Particularly in this case as publication took place almost 110 years after the notes were actually written.

But what does one make of the entry in Doke and Vilakazi (1958: 816) that gives the Zulu word **isi-tingawothi** for stinkwood tree(s)? They do not say which of the various stinkwoods have this Zulu name and indeed there seems no reason why any of the indigenous stinkwoods should be known in Zulu as *isitingawothi* when the various species already have their own genuine, original Zulu names. The Black Stinkwood (*Ocotea bullata*), for example, has the Zulu names **umnukani** (literally, smelling of what?) and the White Stinkwood (*Celtis africana*) has the Zulu name **umvumvu**. Both **umvumvu** and **umnukani** are commonly and regularly used by Zulu speakers. I conclude here that the Zulu word *isitingawothi*, so clearly an adopted version of Stinkwood (or, more likely, the Afrikaans *stinkhout*) refers not to a species of tree, but a type of very well-known and highly useful <u>timber</u>. The timber of various Stinkwoods, like that of the Yellowwood family, is sought after for furniture and I can imagine that *isitingawothi* is a word used by Zulu speakers in the lumber industry and in furniture manufacturing.

The same may well apply to the Zulu word **uloyiphela**, given by Boon (2010: 378) as one of three Zulu names for the Red Pear (or Red Thorn-Pear, *Scolopia mundii*). The Zulu **uloyiphela** is clearly an adopted version of the Afrikaans *rooipeer* (this time with both the initial and the final /r/ of the Afrikaans word becoming /l/ in Zulu). Why does Zulu need to adapt an Afrikaans word for this tree, when the Zulu names **ihlambahlale** and **ingqumuza** already exist? As with the stinkwoods, we note that the wood of the Red Pear is "hard, heavy, used for furniture and wagons" (Boon 2010: 378).

Considerably more problematic are the Zulu names **uloselina** and **uroselina** which Hutchings (1996: 106) gives for *Cinnamomum camphora*, the introduced Camphor Tree. The latter version is clearly the English female name 'Roselina' and the former the same with the initial /r/ rendered as /l/. Boon gives no Zulu names for this tree and Smith (1966) recognises neither 'Roselina' or 'Loselina' as vernacular plant names.

Plant names adopted from Zulu and other African languages

The Umzimbeet and the Tamboetie, the Lala Palm, the Ifafa Lily, madoombies and matoengoelas: these are all well-known and widely used vernacular plant names used by English and Afrikaans speakers in KwaZulu-Natal and the wider South Africa. Less widely used, perhaps, but recorded by Smith in his list of South African vernacular names for plants, are *Umfufu Grass* and *Ipopo Grass*, the *Umtiza Tree*, the *Umthongwani Tree*, the *Umquaqua* and *Umquenque* trees, and the *Isona weed*. All these names have their origins in Zulu or closely related Bantu languages. Let us look at some of them in detail.

The Umzimbeet

The tree *Millettia grandis* is known variously (depending on the writer) as 'umzimbeet', 'umzimbiet', 'omsambeet' and 'umSimbeet' with a variety of other similar spellings. Smith tells us that "Drége in 1830 recorded the vernacular name of *Millettia grandis* as 'Omzambeet', which means 'ironwood'" (1996: 474). Indeed, it does: the Zulu word for this tree, **um-simbithi**, is a compound of the words (*in*)*simbi* (iron) and (*umu*)*thi* (tree). This tree name and other tree names ending with the element -*thi* (tree) are discussed in greater detail in Chapter 3.

The Tamboetie

The tree *Spirostachys africana* is known to Zulus as **umthombothi**, a name I interpret as being a compound of the Zulu words *umthombo* (spring or fountain) with (*umu*)*thi* (tree), but which Nicolaas van Warmelo (1976: 93) intriguingly suggests uses an obsolete Zulu word for stone with (*umu*)*thi*. Again, this is discussed in greater detail in Chapter 3. Boon (2010: 256) gives the English vernacular name as Tamboti and the Afrikaans as *tambotie*. As with the Umzimbeet, there are many spelling variations, such as *tambootie*, *tomboti*, *tamboetie* and others. Smith gives "tambotieboom (-hout) (-wood). Spirostachys Africana" and then says, "see the more correct TOMBOTIBOOM" (1966: 456). For *tombotieboom* he says, "The vernacular prefix [?] 'tombotie' is corrupted from the original native name 'umThombothi', which refers to the caustic properties of the sap and means 'a poison tree'" (1966: 464). There appears to be no evidence to support this interpretation.

The Lala Palm

Hyphaene coriacea is one of the most important trees in Zulu culture and traditional economy as the sap is used to make the popular intoxicating *ubusulu* (palm wine) and no other species of plant is as widely used in the making of baskets and other woven products. The Zulu name **ilala** is not as productive in vernacular spelling variations as **umthombothi** and **umsimbithi** above and Boon (2010: 52) simply gives Lala Palm as the English vernacular name and *lalapalm* as the Afrikaans.

The Ifafa Lily

This is an interesting example as the Zulu-based English vernacular for *Cyrtanthus mackenii* is not an adaptation of the Zulu name for the plant, but the Zulu name for the place where the plant was first found (Smith 1966: 253). The Ifafa Lily (Afrik. *ifafalelie*) is named for the KwaZulu-Natal South Coast district and river named Ifafa. According to Doke and Vilakazi (1958: 197) this place name, as well as the nearby area named Lufafa, is derived from a reduplication of the Zulu ideophone *fá* (sprinkling, drizzling), indicating a place with exceptional mist and drizzle.

Madoombies and matoengoelas

The Zulu word **idumbe** refers to the popular edible tuber *Colocasia esculenta* (Smith 1966: 60; *Arum esculentum*). In the plural this is **amadumbe** and when this word was adapted to the English (and Afrikaans) 'madoombi', the truncated plural prefix *ma-* became part of the vernacular names, requiring a new plural suffix *-s*.

The same applies to the coastal shrub *Carissa macrocarpa* with its delicious vitamin C rich fruits, which can be made into jam. The Zulu word for this shrub is **umthungulu** and **amathungulu** (the same noun stem, but in noun class 6) refers to the fruit of this plant. The Zulu word for the fruit has been transferred to the vernacular name for the species and, as with 'madoombies' above, incorporates the Zulu plural prefix (*a*)*ma-*. Boon (2010: 496) gives Amatungulu as the English vernacular name (with the alternate Big Num-Num), the Afrikaans as *grootnoemnoem*, and correctly says the Zulu name is **umthungulu**.

The Umtiza, the Iboza and various other plant species

The spiny shrub *Umtiza listeriana* has the same vernacular name, *umtiza*, in English, Afrikaans and Xhosa (from which the name originates). There is no Zulu name as the shrub does not occur in Zulu-speaking areas.

Another plant with both the genus name and the English vernacular name taken from the African name is *Iboza riparia*: both Smith (1966: 253) and Pooley (1998: 472) give the English name as Iboza with Pooley supplying the alternate English name Misty Plume Bush. Her botanical name for this species is *Tetradenia riparia*, with *Iboza riparia* given as an earlier synonym. Doke and Vilakazi (1958: 87) give the botanical name of this plant as **Moschosma riparia*, presumably an even earlier synonym.

Both Hutchings (1996: 105) and Smith (1966: 153) give the name 'bog-a-bog' as an English (?) vernacular name for the Lemonwood (*Xymalos monospora*), apparently from the Zulu 'bokoboko' with Smith saying that this is a "Natal name which is unexplained but possibly corrupted from some Bantu name for the species". I am unable, however, to trace the suggested Zulu origin.

A particularly interesting English vernacular name is Panda Tree. Hutchings (1996: 142) gives it as one of a number of English names for *Lonchocarpus capassa*. Clearly this is an adaptation of the Zulu name **umphanda** (Doke and Vilakazi 1958: 645). Smith is not quite accurate when he says that "the vernacular name is derived from the native name 'UMBANDA'" (1966: 363). It is worth noting here that the name of Shaka's half-brother, King Mpande, was frequently recorded as 'Panda' in the writings of earlier explorers and settlers in KwaZulu-Natal. King Mpande and other Zulu kings of the nineteenth century make their way into other vernacular plant names. For example, Hutchings' entry (1996: 238) for the tree *Strychnos decussata* gives among other English names Chaka's Wood and Panda's Walking Stick Tree. Of this same tree Smith says, "Harvey in 1868 described it as *S. baculum* (*baculus* = a staff or walking stick) from the report that the native Chief Panda had his regal staffs cut from the tree" (1966: 363) and Coates Palgrave echoes this history when he says of *Strychnos decussata* that "in the 19th century [sticks from this tree] . . . provided Zulu chiefs with their sticks of ceremony and gave the plant its local name of 'king's tree'" (1977: 765).

A wonderful mixture of alternate names in a variety of languages comes to light when we look into Smith's entry for *impaladoring*. He gives this as a vernacular name for *Acacia goeringii*, saying that "the vernacular name is derived from the fondness of the Impala or Rooibok for the pods, which are said to be very fattening" (1966: 254). He also offers the vernacular names *sambreeldoring* and Umbrella Thorn for this tree. For Coates Palgrave (1977: 242) *Acacia goeringii* is an earlier synonym for *Acacia luederitzii* and the vernacular names are Bastard Umbrella Thorn, *basterhaak-en-steek*, Fat-Thorned Acacia and *buikdoring*. In the erstwhile Rhodesia, the tree was known as the Kalahari Sand Acacia. For Boon (2010: 180) exactly the same species is *Acacia luederitzii* var. *retinens*, with the English name Balloon Thorn, the Afrikaans names *buikdoring* and *blaasdoring*, and the Zulu names **ugagu**, **umshangwe** and **umbambampala**. This last-mentioned name translates as 'what catches the impala' (a reference to the tree's thorns) and so brings us back to Smith's entry *impaladoring* (impala thorn).

Not giving anything like the variety of names recorded above for the *impaladoring*, but providing an intriguing history of the movement of a Zulu name from one name category to another is Smith's entry for **isipingo**:[14]

> **isipingo**. *Scutia myrtina*. The name of the south coast town Isipingo in Natal is said to perpetuate the name (1966: 256).

Doke and Vilakazi (1958: 663) give **isiphingo** as the Zulu name for the thorny shrub *Sentia indica*[15] and indicate that the isiPhingo River south of Durban takes its name from this plant. It is from this river that the South Coast town, now simply a suburb of Durban, takes its name. If I understand Smith correctly here, the English vernacular name 'isipingo' is not an adaptation of the Zulu name *isiphingo* for the same plant, but is taken from the name of the town that, as we have noted, comes from the name of the river, which in turn comes from the name of the plant! Boon, incidentally (2010: 336, 542), gives **isiphingo** as the Zulu name for *Scutia myrtina* (for which he gives the English name as Cat-Thorn) and for *Canthium inerme* (English name: Turkey-Berry).

What I have tried to show is that the botanical names of plants, and the variety of vernacular names for the same plants in various languages, are not to be seen as discrete entities existing in isolation from one another. They are in many cases interlocked, often derived from one another, a translation of another name for the same species or different words with the same underlying meanings. This kind of inter-relationship between different types of plant names will be encountered several times in this book.

---

14. Given as an English, not a Zulu, common name.
15. The genus name *Sentia* is not recognised by Van Wyk et al. (2011).

2 The nature and history of plant naming

> I love their names – lady's bedstraw and red campion, Queen Anne's lace and herb Paris – each one a link with countrymen of the past, who gave them their strange-sounding monikers. Schooled in botanical Latin, I know them also as *Paris quadrifolia* and *Arum maculatum*. But it is the mystery bound up in their common names that soothes me now (Titchmarsh 2008: 325).

In this chapter, I attempt to situate Zulu plant names within a wider historical context. Much of this chapter will look at the contrasts between two different kinds of naming systems, both referred to in the brief quotation above from Alan Titchmarsh's short article 'Heaven in a Wild Flower' (2008). The names *Paris quadrifolia* and *Arum maculatum* belong to "botanical Latin", linked in this quotation to the word "schooled", suggesting that these names have to do with formal education, with scientific theory and knowledge. The names "lady's bedstraw and red campion, Queen Anne's lace and herb Paris" are 'common names', linked to the words 'soothe' and 'love'; words of emotion, rather than science. They have historical value, too, and "link with countrymen of the past".

Titchmarsh talks of the "mystery bound up in their common names". If I read between the lines here, I think his mystery is <u>why</u> certain plants should be called "lady's bedstraw, red campion, Queen Anne's lace and herb Paris". What <u>is</u> 'lady's bedstraw' and how did it come to refer to a plant? What is 'campion'? Why 'Queen Anne's lace', and not someone else's lace? And is this name a metaphor, or does it refer to a historical incident? Why 'herb Paris' and how does 'Paris' get into an English plant name?

But to me there is much more mystery in plant names, not just in the common names, but when various common names and scientific Latin botanical names are put together. One of the more recent books on plants is Richard Boon's *Pooley's Trees of Eastern South Africa*. I open a page at random, find two trees and give their names as set forth by Boon:

Tree One: *Sclerocroton integerrimum* [= *Sapium integerrimum*] – **Duiker-berry**; duikerbessie (A), umdlampunzi, umhlepha (Z); umhlalampunzi (S); amondoli (T).
Tree Two: *Shirakiopsis elliptica* [= *Sapium ellipticum*] – **Jumping-seed Tree**; springsaadboom (A); umdlampunzi (Z); umhlepha (Z, S); umbongolo, umbengele, umwongolo (X) (2010: 254).

We all know that the Latin names are definitive. They identify without question. So from them we see clearly that not only do we have two distinct species of tree here, but two distinct genera: one belongs to the genus *Sclerocroton*, the other to the genus *Shirakiopsis*. And yet *Sclerocroton integerrimus* has in brackets afterwards [= *Sapium integerrimum*] and *Shirakiopsis elliptica* is followed by [= *Sapium ellipticum*]. A botanist will understand that these names in square brackets refer to earlier classification when these two trees were seen as belonging to the same genus. In other words, the names in brackets are earlier synonyms of current valid names. To someone with no botanical training, however, this ongoing process of re-classifying and consequent renaming of plants is mysterious.

And then we come to the English and Afrikaans common names: clearly 'Duiker-Berry' is a translation of '*duikerbessie*'. Or is it the other way around? Is '*duikerbessie*' not perhaps a translation of 'Duiker-Berry'? Which came first, the English or the Afrikaans name? Mysterious. And there is exactly the same problem with 'Jumping-Seed Tree' and '*springsaadboom*'. Which came first?

The biggest problems, however, arise with the names recorded in Zulu, Swazi and Xhosa, all closely related Nguni languages. For *Sclerocroton integerrimum* [= *integerrimus*], Zulu has **umdlampunzi** (what the duiker eats), which seems to tie in with the English and Afrikaans names, but Swazi has '*umhlalampunzi*' (where the duiker lies). Why?

Zulu has '**umdlampunzi**' and '**umhlepha**' for both trees. Why? They are not the same species, or even (today at least) the same genus. Do they both look the same to the indigenous botanist? If we look at Boon's descriptions, one is a "small to medium tree" and the other a "medium to tall tree". One has "smooth, pale-grey bark", the other "rough, longitudinally-furrowed, dark" bark. They can't be that similar. To continue, Zulu uses the word '**umhlepha**' for both species, Swazi only for *Shirakiopsis elliptica*. Why?

And, finally, Xhosa names. I do understand why there are Xhosa names for *Shirakiopsis elliptica*, but not for *Sclerocroton integerrimus* as the distribution maps in Boon show clearly that *Sclerocroton integerrimus* only occurs in Zulu-speaking areas. But why are there three Xhosa names for *Shirakiopsis elliptica* when the distribution map shows it occurs only in a very small part of the Xhosa-speaking area, just south of the KwaZulu-Natal border? And those Xhosa names are suspiciously similar: are '*umbongolo*' and '*umbengele*' not just different interpretations of not-very-clear handwriting? And '*umwongolo*' (*umvongolo*?) and '*umbongolo*' look much like the pairs of names found commonly in Boon's (and others') lists of Zulu names, distinguished only by the contrast between the phonemes /b/ and /v/ that are often interchangeable or indistinguishable in many languages. So many questions and they all arise from the listed names of only two species of trees.

Pavord, in her highly readable history of botanical nomenclature, gives several examples of confusion arising from plant nomenclature. In a reference to 'Juliana's book',[1] Pavord states that it:

> shows a plant with dramatic arrow-shaped leaves, the plant we know as wild arum. Its official name is *Arum maculatum*, a name – thanks to the International Code of Nomenclature – now recognized in Europe as easily as in America, Japan, Australia or India. But when Dioscorides first wrote about this plant it may have been called 'aron', 'aris', 'eparis', 'ephialton', 'kynozolon, 'onekophalon', 'parnopogonon' or 'phoinikion'. If you were Egyptian, you'd have called it 'ebron'; the Romans knew it as 'beta leoprina', the Etruscans as 'gigarum', the Daker as 'kurionnekum', the North Africans as 'ateirnochlam', the Syrians as 'lupha'.[2] If you could not be absolutely sure that this Greek plant was that Spanish (or Italian, or French or Czech, or Polish or German or English) plant, none of the directions for preparing it or

---

1. Around AD 512, the people of Constantinople presented "their imperial patron, Juliana Anicia" with a beautifully illustrated book containing all known botanical knowledge to that date (Pavord 2005: 82).
2. And the Zulus as **umfana kaNozihlanjana**, a name discussed in great detail below.

using it to quell fevers, heal wounds, knit bones, ward off monsters, was of the slightest use (2005: 96).

More than 1 000 years later, between 1652 and 1655, the English scholar John Goodyer wrote:

Iris [somme call it Iris Illyrica, somme Theklpida, somme Urania, somme Catharon, somme Thaumastos, the Romanes call it Radix Marica, somme Gladiolus, somme Opertritis, somme Conseatix, the Egyptians call it Nar] is soe named from the resemblance of the rainbow in heaven . . . Whence for the varietie of colours it is likened to the heavenly rainebow (quoted in Pavord 2005: 74).

Part of the confusion arises, of course, from names in different languages, from different cultures. We could hardly expect the ancient Syrians and Etruscans, the Egyptians and Romans, the thirteenth-century English and French, to have the same names for certain plants. But surely within one language, common names for plants have not provided mystery and confusion? Unfortunately, this is certainly the case. Pavord tells us of the problems the Greek scholar Theophrastus (*c.* BC 372–287) had with regional variance in vernacular names:

'the Arcadians appear to differ as to the names which they give'. Painstakingly, Theophrastus picks his way through this muddle of names . . . calmly laying out the facts, highlighting the areas where further enquiry is needed. In the mountains a certain maple is called *zygia*. In the plains, it is *gleinos*. Is this the same tree, under different names, or two different kinds of maple? (2005: 31)

The problem of multiple names for single plants does not improve over the centuries. In fact, one might say that the problem multiplies exponentially with time. Pavord gets from William Stearn's *Dictionary of Plant Names for Gardeners* (1996) the following information:

A widespread flower such as the marsh marigold (*Caltha palustris*) has about sixty common names in France, another eighty in Britain, and at least 140 in Germany, Austria and Switzerland (2005: 396).

In the face of this colossal multiplicity, the seven Zulu names for the common Blackjack Weed (*Bidens pilosa*) and the eleven names for *Trema orientalis* (Pigeonwood) pale into insignificance. Here are further examples of multiple names for a single plant in a single language:

> Dog's Mercury (*Mercurialis perennis* Linn., adder's meat, boggart flower, boggart posy, dog flower, dog's medicine, green waves, snake's bit, snake's flower, snake's food, snake's meat, snake's victuals, or just simply snakeweed) is used in herbal medicine for enemas, but otherwise has all the characteristics that are summed up by the herbals: it is 'a hairy, creeping, foetid perennial', and it also tends to colour things yellow (McLaughlin 1971: 78).

> *Lantana* is the ancient name of Viburnum, a European plant that has similar foliage . . . In other countries *Lantana* is called wild oregano, cherrypie, English sagebrush, confetti bush, red sagebush, red horse-tassel, tickberry and Jamaican mountain sage (Stirton 1978: 91).

Pavord's book traces mankind's search for order in the naming of plants. As she points out several times, for much of the last 2 500 years, the study of plants and the study of medicine went hand in hand. Writing about the relationship between plants and medicine in early seventeenth-century London, she states:

> The apothecaries have a particular need to sort, name and categorise indigenous plants. Medicine is their business and plants the raw material from which they brew, distil and decoct their elixirs and tonics (2005: 5).

As we saw above in the quotation from 'Juliana's book' on the identification of the Wild Arum, if an apothecary or doctor could not be absolutely sure of the identity of a plant, then "none of the directions for preparing it or using it to quell fevers, heal wounds, knit bones, ward off monsters, was of the slightest use". And again:

> and yet [plants] must have names, if only for practical reasons . . . A plant's pharmaceutical value depended on the plant-hunter's ability to distinguish one botanical species from another . . . (2005: 14).

The study of plants was . . . intimately connected with the study of medicine. A sixteenth-century apothecary, surgeon or doctor had necessarily to be a plantsman (2005:16).

Pavord's book shows how, over the centuries, information about plants and their uses accumulated and was shared as trade routes and exploration opened up the world more and more. Unfortunately, the people working with plants were from a wide variety of cultures and linguistic backgrounds: the more botanical knowledge accumulated the more chaotic and muddled the naming became. Pavord quotes the Italian Brasavola of Ferrara, who writes in 1536 about a "plant he calls *Primula veris*":

This is what your apothecaries take to be *Primula veris*, by our women called *petrella*, by some other St Peter's herb. Among recent names are *herba paralysis* and *margarita*, so much are people given to imposing names each according to his own fancy (2005: 216).

Pavord comments about this as follows:

He's talking about a marguerite, a daisy, more specifically the common daisy *Bellis perennis* . . . This plant had been growing in Europe for as long as people had inhabited it, yet still there was no consensus about its identity. The key lay in providing universally accepted tags to set alongside the common names of plants. The common names, of course, differed from country to country; even within one country, as Brasavola demonstrated, four or five different names might be attached to the same plant. They still are (2005: 218).

It became more and more clear that some sort of universally understood system of nomenclature had to be adopted and the ground-breaking work for this was mainly achieved by the efforts of the English 'plantsman' John Ray (1627–1705) "whose *Synopsis methodica* of 1690 laid down the rules for a modern system of nomenclature" (Pavord 2005: caption to plate 147). According to Pavord, the "six Rules Ray proposed provided the vital underpinning of a new discipline which would later acquire a new name – taxonomy" (2005: 392).

Pavord's exhaustive history of the search for a universal system of nomenclature ends with John Ray, leaving the average reader to wonder where the well-known scholar Carl Linnaeus of Sweden fits into the picture. For this, the reader must wait for Pavord's epilogue, which begins:

Of course, the story does not end with Ray. He forges the rules that will steer his successors through the complex maze of nomenclature that lies ahead. He establishes the study of plants as a scientific discipline. He gives this study a new name – botany (2005: 395).

And then, later on the same page, Linnaeus finally makes his appearance:

And somewhere we have to nod, however grudgingly, to Carl Linnaeus (1707–1778), the Swedish botanist who described his own book, *Species plantarum*, published in 1753, as 'the greatest achievement in the realm of science'. Enthroned as professor of medicine at the University of Uppsala, he called his students 'apostles'. Like Mattioli, he had the good fortune to publish the right book at the right time. He captured the *zeitgeist*, understood what was required and, with the ruthless efficiency of a computer programme, imposed two-module name tags on nearly 6,000 plants (2005: 395).

Most writers on botanical nomenclature seem happy to give all the credit for the modern system of botanical nomenclature to Linnaeus. Charles Heiser ascribes the Latin name of maize (*Zea mays*) to "Linnaeus, the great Swedish botanist of the eighteenth century" (1973: 99). In Liberty Bailey's book *How Plants Get Their Names*, the opening line of the chapter headed quite simply 'Linnaeus' reads, "It is profitless to go further in quest of names until we know Linnaeus" (1963: 16). None of Pavord's dozens and dozens of earlier scholars of plants and plant naming is mentioned. Bailey simply says that the system of binomial nomenclature "begins with Linnaeus in Species Plantarum[3] in 1753: that date is the starting-point for the naming of plants" (1963: 24). However, it is acknowledged that:

We must not conclude from the foregoing discussion that two-word designation of plants was unknown before Linnaeus. Open on my table is a choice vellum book of the Frenchman Carolus Clusius . . . [P]rinted in 1576 . . . [H]ere is a picture named Genista tinctoria, another entitled Dorycnium Hispanicum, and many others. These names were not part of an organized system, however (1963: 24).

---

3. Bailey gives this book title without italics.

Later, Bailey is back with open admiration of Linnaeus:

> Thus, now, have we made brief acquaintance with Linnaeus, sometimes known to moderns as the 'father of botany' . . . His was a systematic synthetic mind. He united the scattered essentially unclassified records of centuries. He brought order into the study of plants (1963: 30).

This echoes a remark of the Linnaeus-resistant Pavord that:

> The brilliant glory lily, which created a sensation when it was first brought into Europe from the tropics, had been 'Methonica' to one nursery, 'Lilium zeylanicum superbum' to another, 'Mendoni' to a third. Linnaeus decreed that it should henceforth be called *Gloriosa superba*, one of its earliest names. And, surprisingly, the rest of the world eventually agreed. Just in time, order had been wrested from chaos (2005: 396).

One wonders just how much this last remark is tongue-in-cheek. So, thanks to Linnaeus, since 1753 we have had order rather than chaos in the world of botanical nomenclature. Not everyone necessarily agrees with that and certainly after I have spent many fruitless hours attempting to correlate the Latin names (earlier synonyms) given in Doke and Vilakazi's *Zulu-English Dictionary* (1958) with the Latin names in botanical books of a later era, I have often wished for someone to bring order out of chaos.

Indigenous knowledge versus scientific botany

We now go on to investigate the major differences between the system of common names for plants ('chaos') and the system of scientific Latin-based binomial botanical nomenclature ('order'). Effectively, we are comparing scientific botanical knowledge with what has come to be known in South Africa at least as indigenous knowledge (as in indigenous knowledge systems or IKS).[4] It helps to see the contrasts in the form of a table:

---

4. Elina Helander-Renvall (2007), in her article on the relation between snow and reindeer herding among the Sámi people of the Nordic countries, uses the term 'traditional ecological knowledge'.

Indigenous knowledge	Scientific botany
'Common' or vernacular names	Latin names ('scientific' or 'botanical')
Naming 'just happens'	Naming is deliberate, intentional, planned
Namers unknown	Namers are identified, known individuals
Date of naming unknown	Date of naming and identification are recorded
Meaning very often entirely referential	Referential and underlying meanings
Unsystematic	Systematic
Range from single words to phrase	Basically binomial
Names regional, linked to specific languages and cultures	Names 'universal'
Knowledge passed on orally	Knowledge passed on academically
Knowledge stored in group memories	Knowledge stored in written form

Let us take each of these contrasting sets in turn.

Naming 'just happens' versus deliberate, planned naming

In a quotation from Pavord about the scientific name of the "brilliant glory lily", we learn that "Linnaeus decreed that it should henceforth be called *Gloriosa superba*, one of its earliest names" (2005: 396). We learn also that Linnaeus imposed binomial name tags on nearly 6 000 plants, these tags comprising a genus name plus species name or epithet. This is deliberate, planned naming. In contrast, the common names of plants, in whatever language, 'just happen'. No one can explain why an oak is an 'oak', or a willow a 'willow'. We can, of course, investigate the etymologies of names like 'oak', 'willow', 'daisy' and 'rose', and discover that they are Old English, adopted from West Saxon, mutations of Old Norse words or entered English from Latin via Old French. But this does not help us understand why the West Saxons, early Norsemen, Greeks or Romans first chose a particular word for a particular plant. The words 'just happened'.

Namers unknown versus namers known

This contrast is similar to the set above. We know that Linnaeus named 6 000 plants. Pavord's book is filled with examples of early, named scholars naming this plant and that. In modern botany textbooks, many plant descriptions include the name of the person who named the plant. Bailey offers the example "Sassafras was named *Laurus variifolia* by Salisbury in 1796" (1963: 69). Not only is the namer (in botany called the 'author' of the name) named, but also the publisher. Bailey points out:

> If the indication of the binomial is to be accurate and complete, and in order to verify the date, it is necessary to quote the author who first published the name: *Parthenocissus quinquefolia*, Planch (1963: 71–72).

Compare this situation with that of common names. Who first called a myrtle 'myrtle'? Who was the originator of the word 'lily'? What individual or individuals is or are responsible for the words 'beech', 'pine', 'sorrel' and 'plantain'? These questions are unanswerable.

Date of naming unknown versus date of naming and identification recorded

See the quotations above for naming and dates of naming by the Latin botanical name. The only dates we have for common names are in etymological dictionaries, in which information is given on the approximate date a particular word or name was used in the language, as for example 'larch' (sixteenth century from German *lärch*), 'chestnut' (sixteenth century from earlier *chesten nut* < Old French *chastaigne*), 'bush' (thirteenth century of Germanic origin), 'teak' (seventeenth century from Port. *teca* < Malayalam *tecca*), and so on. A few Zulu common names adopted from other languages can reliably (but not very accurately) be dated after European settlement. Examples are **umsilinga** (< *seringa*) and **uloyiphela** (< Afrik. *rooi peer*, Red Pear). But as for giving the actual year in which a common name was first used for a plant, clearly this is quite impossible.

Meaning very often entirely referential versus referential and underlying meanings

Chapter 4 goes into detail about different types and layers of meaning. Here I can give only the briefest summary. The primary function of all names, of whatever kind, is to identify, denote, refer to an entity, whether person, place or plant. It follows then that the primary meaning of a name is referential, or denotational. Many names have other layers of meaning. The name 'sunflower' not only refers to a plant known by this name, but also tells us that this is a flower linked to the sun. Such underlying meanings are not always overt. The name 'daisy' hides the Old English form *dægesege* (day's eye), a reference to the fact that daisies commonly open during daylight hours only. A similar covert meaning is found in 'dandelion', a fifteenth-century English adoptive from Old French *dent de lion* (tooth of a lion) in reference to its leaves.[5] The derivation of 'tulip' refers to the shape of the flower: it is derived from New Latin *tulipa* from the Turkish *tulbend* (turban).

Far more hidden, though, is the underlying meaning in 'celandine' (earlier celydon) derived, via Latin *chelidonia*, from the Greek *khelidon* (swallow) in the belief that the plant's seasons paralleled the migration of swallows (*Collins Dictionary of the English Language* 1986: 254). One searches in vain, however, for similar meanings in acacia, almond, beech, birch, bryony, cedar, clover, cypress, daffodil, elder, elm, fig, hazel, hemlock, holly, juniper, lavender, lily, mahogany, nettle, oak, pansy, pine, plum, teak, thistle, violet, yarrow, and hundreds of others. These names just 'are'.

If we compare these common names, by far the majority of which have the single function and single meaning 'referential', to Latin botanical names, we note that each and every botanical name has at least two functions, and at least two meanings. Again, I stress that the primary function of any name (and so the primary meaning) is referential. The primary function of a name like *Gladiolus papilio*, a widespread South African flower of marsh and damp grassland, is to identify this flower. The genus name *Gladiolus* separates this flower from all other plants in the world not found in this genus, while the specific name *papilio* separates it from the 25 other members of the gladiolus genus found in South Africa (for example, *Gladiolus auranticus*, *G. crassifolius*, *G. cruentus*, *G. dalenii* and several others) (Pooley 1998: 607). But the name has a separate function with related meaning: *Gladiolus* is Latin for 'small sword' and refers to the leaf shape of this genus; while *papilio* means 'butterfly-like', a reference to the flowers as also seen in the English common name Butterfly Gladiolus.

Latin botanical names have three distinct categories of secondary meanings:

Physical description
Indigofera (bearing indigo), *eriocarpa* (woolly fruits)
Tephrosia (ashen, grey-green/silvery leaves), *elongata* (elongated)
Eriosema (woolly petal), *distinctum* (separate)

---

5. Beaumont explains in a personal communication that the "leaf outline is deeply indented with the appearance of the outline as a series of sharply pointed teeth".

Locational

Tricliceras mossambicense (from Mozambique)
Androcymbium natalense (from Natal)
Selago monticola (growing on hills)

Honorific or commemorative

Schizoglossum hilliardiae (named after Olive Hilliard, South African
  botanist)
Sparmannia ricinocarpa (named after Swedish botanist Andrew
  Sparrmann, 1748–1820)[6]
Helixanthera woodii (named after John Medley Wood, 1827–1915,
  English-born South African botanist and founding curator of the Natal
  Herbarium)

Unsystematic versus systematic

This is a debatable contrast. Common names may not be systematic in the sense
that Latin botanical names are, but can, it has been argued, be systematic in
a different kind of way. Louis Louwrens, at one time professor of Northern
Sotho at the University of South Africa (UNISA), argues (1999) for hierarchical
categories of vernacular plant names. His description of Northern Sotho plant
taxonomy is based on work by Berlin (1976) and Berlin, Breedlove and Raven
(1973), and presents a hierarchy of five taxa: unique beginner (e.g., 'plant'),
life form (e.g., 'tree'), generic (e.g., 'bushwillow'), specific (e.g., 'velvet
bushwillow') and varietal (no example given). His description of Northern
Sotho plant names using these taxa makes it clear that a system is operating.

The system of Latin-based botanical nomenclature is of a different nature.
Honed by expert botanists, horticulturalists, herbalists and professors of both
medicine and botany over 300 years, this system is ratified by international
botanical congresses. For example, the first International Botanical Congress
(IBC) was held in Vienna in 1905 and the second IBC in Brussels in 1910
(Bailey 1963: 74). Extensive rules are set up regarding descriptions, authorities,
precedence, use of Latin and much more. These rules are set out in the
International Code of Botanical Nomenclature, which is constantly revised.

---

6. Beaumont confirms (personal communication) that the man's name is Spar̲r̲mann, but
   the genus name is *Spar̲mannia*.

One could, in the context of plant naming, talk of <u>three</u> different systems. For most people, 'systematic plant naming' means the Latin-based binomial system popularised by Linnaeus. Louwrens (1999), following Berlin, Breedlove and Raven (1973), argues convincingly for a system of 'folk botanical nomenclature' and his description of Northern Sotho plant names holds good for Zulu plant names, as we shall see later. But in fact all naming, of whatever kind, takes place within a system, in which there are four major categories: names; the named; namers; and the context in which naming takes place. These four categories themselves occur in hierarchical sub-categories with inter-relational dynamics taking place at all levels. This must be considered a third, and different, kind of system.

Range from single words to phrase versus binomial system

The very description of the term 'binomial' in reference to Latin-based botanical nomenclature explains that these names always occur as a two-word phrase, of which the first is the genus name and the second refers to the individual species within each genus. Occasionally, very occasionally, botanical works offer names that occur as three-word phrases: *Solanum maniacum multis* (Bailey 1963: 102) and *Leucadendron spissiflium* subsp. *natalense* (Pooley 1998: 248), for instance.[7]

The two-word phrase, however, is only one possibility for vernacular names, and by no means the most common. Common names occur perhaps most often as single words. Earlier, I gave a list of common plant names in English that ran from acacia to yarrow. Afrikaans, too, has many single-word plant names, such as *lelie* (lily), *klawer* (clover), *tulp* (tulip), *roos* (rose), *affodil* (daffodil),[8] *laventel* (lavender), *kastaing* (chestnut), *seder* (cedar), *lourier* (bay), *sipres* or *sapree* (cypress), *lork* or *lariks* (larch), *den* (pine) and *akasia* (acacia). The Afrikaans language, though, prefers to add -*boom* (tree) or -*hout* (wood) to single words for tree, to form another single, but compound, word: *denneboom* instead of *den*, *lariksboom* for *lariks*, *sederboom* or *sederhout* for *seder*, and

---

7. Beaumont, however, states (in a personal communication): "There are no trinomials in botanical nomenclature (although they do exist in Zoology). 'Multis' may therefore be a subspecies (subs) or variety (var.) or a hybrid (x) and any of the abbreviations may be between 'maniacum' and 'multis'."

8. *Collins Dictionary of the English Language* (1986) gives the etymology of 'daffodil' as "from Dutch *de affodil*" (< *asphodel*) so it is not surprising that the Afrikaans is *affodil*.

so on. Most single-word plant names in Afrikaans are compounds as in the following examples from Pooley (1998): *bitterappel* (bitter apple), *wildepatat* (wild potato), *wildestokroos* (wild hollyhock), *voëlbrandewyn* (bird's brandy), *peultjiesbos* (small-pod bush) and *ystervarkwortel* (porcupine root). In Zulu, too, many single-word plant names are compounds and we shall look at these in detail in Chapter 3.

Many common names have more than two elements, as in the following examples. In Stevens (2008) we find Devil's Bit Scabious, Codlins and Cream (also known as Great Hairy Willow-Herb), Lords and Ladies (also known as Cuckoo Pint), Bacon and Eggs (see Bird's-Foot Trefoil), Bird's-Nest Orchid and Old Man's Beard (also known as Traveller's Joy). Boon (2010) gives tree names like Broad-Leaf Waxberry, White Stinkwood (see False White-Stinkwood and Red-Fruit White-Stinkwood), Large-Leaved Rock Fig and Scarlet Silky Oak.

Afrikaans and English vernacular names from Smith (1966) include: *grootrooi-afrikaner* (*Gladiolus aestivalis*), Little-Man-in-the-Boat (*Strelitzia reginae*), Lily-of-the-Nile (*Zantedeschia aethiopica*), *koning-van-kandië* (*Haemanthus coccineus*) and the delightful *juffertjie-roer-by-die-nag* (Little-Lady-Gad-About-At-Night, several species of *Struthiola*).

Knowledge passed on orally versus knowledge passed on academically

Latin-based nomenclature is taught in botany classes at schools and tertiary institutions. Like all formal academic knowledge, such botanical knowledge is linked to structures such as classrooms and lecture theatres, notes and textbooks, formal instruction times, and of course, testing of knowledge in assignments and examinations. Vernacular-based plant nomenclature ('common' names) tends to be passed on orally, although one must acknowledge several books on vernacular names such as Smith (1966) and Stevens (2008).

In traditional Zulu society, botanical knowledge is mainly held in the heads of *izinyanga* (traditional healers, herbalists or medicine men) and it is for this reason that such herbalists and traditional healers were the focus of many hours of interviews for the Zulu Botanical Knowledge Project (ZBKP, of which more at the end of this chapter). The sons (grandsons, nephews) of *izinyanga* are usually apprenticed to the older, knowledgeable doctor and spend many years learning their knowledge.

Knowledge stored in group memories versus knowledge stored in writing

This point is similar to the previous one, but it is introduced under a separate heading simply because when names are recorded in writing there are various possibilities for their rendering not available for the oral form. There are also issues related to the permanence of archiving that provide a contrast between oral and written forms of knowledge.

Let us take the first point using a single Zulu name for a plant as an example, the Zulu name for the Monkey-Tail Everlasting (*Helichrysum herbaceum*) that Pooley (1998: 314) gives as **impepho-yamakhosi**. The Zulu word *impepho* here refers to any species of plant, usually of the *Helichrysum* genus, that in its dried form is burnt as incense in shamanistic ritual. The word *yamakhosi* literally means 'of the chiefs' and is here a synonym for *amadlozi* (ancestral spirits). When this name is elicited from a mother tongue Zulu-speaker, the answer will be the oral utterance represented here by the letters of the written word **impepho-yamakhosi**. The phonemes (speech sounds) represented by these letters will always be the same and always appear in the same order. There is only one oral version of this plant name.

Once this name is written down in a field notebook, however, and later transcribed, typed, entered into a word processor and edited for publication, there is no longer only one potential version. The individual handwriting in the field notes becomes typeface and the person entering the handwritten entry as type must choose from a variety of fonts, font sizes, modes such as italic or bold, and so on. The author or editor must choose whether and where to capitalise and write phrases as a single word, separate words or hyphenated words.

The single oral form represented by Pooley above in the written form **impepho-yamakhosi** could appear in a considerable number of different forms, such as the following:

impepho-yamakosi	impephoyamakhosi	imphepho yamakhosi
iMpepho yamakhosi	Impepho yamakosi	imPepho yamakhosi
iMpepho-yamakhosi	Impepho-yamakosi	imPepho-yamakhosi
iMpepho yaMakhosi	Impepho Yamakhosi	imPepho yamaKhosi
iMpephoyamakhosi	Impephoyamakhosi	imPephoyamakhosi

A comparison of five authors

Authors and editors of botanical publications have to choose how to write plant names. As an example of variation, I look below at the way the names of twelve plants have been rendered in Zulu. The authors are Boon (2010), Pooley (1993), Hutchings (1996), Moll (1981) and Doke and Vilakazi (1958). It should be noted that these five publications differ over more than simple issues of hyphenation and compounding.

Botanical name Hutchings	Boon Moll	Pooley Doke and Vilakazi (D&V)
Sideroxylon inerme umakwela fingqane [*sic*]	umakhwela-fingqane uMakhwela-fingqane	uMakhwela-fingqane umakhwelafingqane
Oxyanthus latifolius not in Hutchings	umaphekemoyeni- omnyama uMaphekemoyeni- omnyama	uMaphekemoyeni-omnyama not in D&V
Ekebergia capensis umathunzini wentaba/ umathunzini-wezintaba	umathunzini-wezintaba uMathunzini-we-zintaba	uMathunzini-we-zintaba umathunzi wentaba
Erythroxylum delagoense not in Hutchings	umbhuletsheni- wangaphandle uMBhuletsheni- wangaphandle	umBhuletsheni-wangaphandle not in D&V
Combretum erythrophyllum umdubu (-wehlanze)	umdubu-wehlanze umDubu	umDubu umdubu wehlanze
Combretum collinum Umbondwe-mhlophe/ umbondwe (-omhlophe)	umbondwe-omhlophe umBondwe-omhlope [*sic*]	umBondwe-omhlophe umbondwe
Bridelia micrantha umhlalamagwababa/ umhlamagwababa [*sic*]	umhlalamagwababa umHlalamagwababa	umHlamagwababa umhlalamagwababa
Chionanthus peglerae not in Hutchings	ithintane-elimhlophe iThintane-elimhlophe	iThintane-elimhlophe ithintane elimhlophe
Trema orientalis isiKhwelamfene	isikhwelamfene isikhwelamfene	isiKhwelamfene/iSkolemfene [*sic*] isikhwelamfene
Lumnitzera racemosa not in Hutchings	isibhaha-esibomvu not in Moll	isiKhaha-esibomvu [*sic*] isibhaha
Encephalartos natalensis Isigqiki-somkhovu	isigqiki-somkhovu isiGqiki-semkhovu [*sic*]	isiGqiki-semkhovu [*sic*] not in D&V
Sideroxylon inerme amasethole (-amhlophe)	amasethole-amhlophe aMasethole-amhlope [*sic*]	aMasethole-amhlope [*sic*] amasethole amhlophe

Correspondence between plant and name

There are three possibilities for correspondence between plant and name:
- One name, one plant
- One name, many plants
- One plant, many names

The first is of course the stated goal of Latin-based botanical nomenclature. Ideally, every species of plant has its own distinct (binomial) name referring to that species and no other. That this is not always the case is simply because of continual reclassification by botanists, or the discovery of earlier names that may take precedence over later, widely known names.

Examples two and three are typical of folk botanical nomenclature. Words like 'daisy', 'iris' or 'lily' may refer to hundreds of different species, requiring them to be distinguished by a variety of different qualifiers. Then again, one species of plant may have many names. And there is a fourth option for plant names. One name may have many different forms, of which many (if not all) are regarded as the correct form by one or other authority.

One name for several plants

The following are a few names that refer to a multiplicity of species. The plant name **idololenkonyane** (calf's knee) is used for the following species of plants: *Cyphostemma natalitium*, *Rumex lanceolatus*, *Persicaria lapathifolia*, *Barleria obtusa*, *Rumex crispus* and *Hydnora africana*. When the name is further qualified by **elimhlophe** (<u>white</u> **idololenkonyane**), it refers to *Hypoestes aristata*.

The name **umlunge** is used for *Crocosmia paniculata*, *Eulophia welwitschii*, *Gladiolus sericeovillosus*, *Watsonia densiflora* and *Dicoma zeyheri*. When **omhlophe** (white) is added, it refers to *Eulophia speciosa*.

Several names for one plant

The Common Blackjack Weed, *Bidens pilosa*, has seven Zulu names according to Pooley: **umhlabangubo, amalenjane, isikhathula, ucucuza, ugamfe, umesisi** and **uqadolo**. Other plants with many Zulu names include: *Salacia kraussii* (with five) – **ibhonsi, umbhonsi, ihelehele, umgunguluzane** and **umnozane**; **Scilla nervosa* (five) – **ingcino, ingcolo, umgcinywana, imbizankulu** and **ingema**; and *Gomphocarpus physocarpus* (also with five) – **umangwazane, umbababa, umqumbuqumbu, uphuphuma** and **usingalwesalukazi**.

Many plants have four names. Two or three names are the norm. The Pigeon Wood (*Trema orientalis*) holds the record for the highest number of Zulu names for one plant. Pooley (1993) and Boon (2010) give a total of nine names and Doke and Vilakazi (1958) add another two, making 11 in all: **isikhwelamfene, ifamu, iphubane, ubathini, umbhangabhanga, umbokhangabokhanga, umcebekhazane** [*sic*]**, umdindwa, umsekeseke, iphumela** and **umphumelele**.

Variations in oral knowledge

As already stated, indigenous knowledge tends to be archived in group memories and manifested orally. It has been noted by scholars such as Ong (1982) that written knowledge is characterised by fixed forms, while oral knowledge is characterised by variant forms. For example, scholars of oral poetry around the world have pointed out that in an oral tradition there is no such thing as a 'single, definitive text'. The praises of King Shaka, recited by one *imbongi* (oral poet) will differ from the 'same' praises recited by another *imbongi*, even though both will (correctly) claim to be familiar with the praises of Shaka. Indeed, the praises recited by one *imbongi* will differ from those recited by the very same *imbongi* a week later. When I say 'differ', I mean from a Euro-Western viewpoint. Version A might be longer or shorter than version B, the order of stanzas will be transposed and certain details may be left out or added. But to a Zulu audience version A will be every bit as authentic as version B, or any other versions that the same audience may hear over the years. Rycroft and Ngcobo say this about transcriptions of oral poetry:

> It must be strongly emphasized . . . that such transcripts do *not* represent fixed, 'standard forms' of these eulogies, but only one out of an infinite number of variants, in each case (1988: 11).

I would like to illustrate this principle with one single plant name, a name I call 'one single plant name' even though four publications produce ten variations of it and I could theoretically predict another 30 or 40 possibilities. Our single name here is **umfanakazihlanjana** (little-boy-of-the-marsh). Cross-referencing between Pooley (1993), Hutchings (1996), Doke and Vilakazi (1958) and Ngwenya, Koopman and Williams (2003) produces the following variations, mostly referring to a group of Arum Lilies of which *Stylochiton natalensis* (Bushveld Arum) is representative:

umfana kazitanjana
umfanakazihlanjana
umfana-kaHlanjana
umfana-kaNozihlanjana
umfana-kaSihlanjana
umfana-hlanjani
umfana-ka'sihlanjana
umfana-kahlanjana
umfana-kahlanjane
umfana-kanozihlanjana

Note that **umfana[-]ka** is common to all except **umfana-hlanjani,** and -**hlanjana** common to all except **umfana kazitanjana** and **umfana-hlanjani**.

Neither the botanical reference works, nor Doke and Vilakazi's dictionary, have attempted to analyse this plant name. Its analysis is not, however, difficult, being a compound (or phrasal) name consisting of *umfana* (boy) + *ka* (of, son of) + -**hlanjana** (diminutive form of the noun stem -**hlambo** or low-lying marshy area). This noun occurs as a class 7 noun **isihlambo**, with its plural in class 8, **izihlambo**, which accounts for the -**zi**- and -**si**- in front of -**hlanjana** in the examples above. In **umfana-kaNozihlanjana** (also without capitals), the name formative -**no**- is pre-prefixed to -**zi**-. This formative is frequently found in the names of species, whether botanical or zoological, and may be regarded as an optional morpheme in that its presence or absence makes no difference to the meaning of a word. The presence or absence of -**no**-, presence or absence of noun class prefix in singular or plural are all allomorphic: they are morphological variations that do not change the meaning of the plant name.

The first version in the list above, **umfana kazitanjana**, I would regard as a 'memory fault' and **umfana-hlanjani** as an even worse example. These two versions are, to my mind, defective. With regard to **umfana-kahlanjane**, it has been shown by a number of scholars that the diminutive suffix -**ana** has the allomorph -**ane**, especially in the names of species, and this can also be regarded as a 'free variation' that does not change meaning.[9]

So much for morphological analysis, but what about 'why?' Why should this plant get such a name? Again, this is not difficult to answer. A great number of Zulu plant names in their underlying meaning make reference either to the

---

9. Rycroft and Ngcobo note of the two spellings 'Dingana' and 'Dingane' that "the distinction is very small and a disputable one" (1988: 1).

appearance of the plant or its customary habitat. Arum lilies are normally found in low-lying marshy ground, so the habitat reference is clear. But there is also a likely reference to appearance and I am grateful to my wife Jewel Koopman for pointing out that the projecting yellow spadix of the Arum flower could be the little 'boy' referred to in the name, or even more likely to the little penis of the 'boy of the marshes'. The perceptive metaphor, personification and underlying humour are all very typical of Zulu folk poetry and naming.

Of the ten variants of the name given above, I have indicated that I regard two as defective, but the remaining eight are all possible correct versions of this name. One could go much further than this. Working on the basis that the underlying meaning of the name is 'boy of the marsh', the only basic elements we need are *umfana* (boy), *-ka-* (of) and the noun stem *-hlambo* (marsh). The formative *-no-* can be present or absent; the noun can be given in the singular form *isihlambo* or the plural form *izihlambo*; the stem can be in the diminutive form *-hlanjana* or not; and the diminutive suffix can occur as *-ana* or *-ane*. None of these differences makes the slightest difference to the meaning of the name or to a perceived 'correctness' in oral-based knowledge. Let us compute the possibilities:

- Base form: '*umfana-kahlambo*';
- Presence or absence of *-no-*: two possibilities;
- Singular or plural stem: four possibilities;
- Diminutive or not: eight possibilities;
- *-ana* or *-ane*: sixteen possibilities.

Every one of these sixteen possible oral forms is correct. Now, to make matters far more complex, the transcriber of these sixteen possible oral forms has more choices. He or she can decide whether to write the oral form as one word (**umfanakanozihlanjana**, say), as a hyphenated phrase (**umfana-kanozihlanjana**) or as separate words (**umfana kanozihlanjana**). This option raises the possibilities to 48. Then there is the matter of capitalisation: does one capitalise the first 'n' of '*nozihlanjana*' or not? This option raises the possibilities to 96. Let us emphasise this point. There are <u>96</u> different ways of writing the Zulu name of the Arum Lily *Stylochiton natalensis* and every one is potentially correct. Furthermore, we note that one of the published versions is **umfana-ka'-sihlanjana**, adding the possibility of a single apostrophe after *-ka-*, and a second hyphen. I do not personally regard either of these as correct written forms, but if we did, the potential spellings of this single plant name would climb to a truly splendid 384 versions.

The Zulu Botanical Knowledge Project

It should be stressed that when several names are given for one plant, or when six or seven plants share six or seven names, apparently on a haphazard basis, these sets of several names, whether for one plant or many, are not simply free or arbitrary versions. When an *inyanga* or a plant collector decides to harvest some *Bidens pilosa* (Common Blackjack Weed), he or she does not simply say, 'Well, today I'm going to use the name **umhlabangubo**, but tomorrow I'll use the name **amalenjane**, or perhaps **isikhathula**, and if I still need more next week, I'll tell people I'm going to collect **ucucuza**, or perhaps **ugamfe**; no, wait – **uqadolo** . . .'

The use of these different names is controlled by a number of factors. Regional variation is one of them. As with bird, mammal and reptile names, identification of species may vary from one district to another. North of the uThukela River, a hippopotamus is *imboma*; to the south it is *imvubu*. The same river divides those who call a Glossy Starling *ikhwezi* from those who use the term *ikhwinsi*. Many plant names are known by one name in one part of the Zulu-speaking region and by another or other names elsewhere. As noted in Chapter 1, both Gerstner (1938, 1939, 1941) and Smith (1966) indicate the region where a particular name is used.

Another factor is what I could loosely call the 'aspect' of the plant. If the person who is identifying the plant by name is thinking about where it grows, one particular name may be used; if the same person is thinking about the overall appearance of the plant, another name may be employed. If the usage of the plant is what the harvester has in mind, a third name may be used. Doke and Vilakazi (1958) have a number of entries in which a plant name gloss contains the words "a herbalist's name for the . . .". My colleague Ndela Ntshangase has, in many interviews, said, 'that name [one of many for the same plant] refers to its use as muti'.

Many Zulu plant names have been recorded in written form. How correct the written records are in reflecting the right plants is a matter of conjecture. But knowledge of which variation to use in a particular setting is almost entirely oral knowledge, passed down from one person to another. It was suspicion that this oral knowledge was gradually being lost that started, in the late 1990s, a project that would eventually become known as the Zulu Botanical Knowledge Project (ZBNP).

The idea began in the late 1990s with Mkhipheni Ngwenya and Rosemary Williams, both of whom were working at the time for the Natal Herbarium at the Botanical Gardens in Durban. The original idea was simply to capture

Zulu oral botanical knowledge before it was lost. A working group was set up and during the following four or so years a group of Zulu linguists and botanists, with one or two information technology botanical experts, met regularly. Funding was accessed from the National Research Foundation, the World Wildlife Fund and the National Botanical Institute. Three areas that were considered to be both botanically and linguistically diverse in KwaZulu-Natal were identified – Bulwer/Lotheni, Nkandla-oNgoye and eNtimbankulu – and Ngwenya, with a few assistants, set out to identify local people with extensive oral plant knowledge and interview them. The initial goal of the project was three years in the field accumulating oral plant knowledge via interviews and simultaneously collecting specimens; one year to enter this knowledge into a database; and then three years for a lexicographical team, including postgraduate students, to collate this information into a Zulu Plant Names Dictionary.

To satisfy the funders, and show that their money was being well spent, it was agreed to produce a short publication after three or so years. The result was *Ulwazi LwamaZulu Ngezimila: Isingeniso/Zulu Botanical Knowledge: An Introduction* (2003), a bilingual 67-page book co-authored by Williams, Ngwenya and this writer.

Sad to relate, due to a shortage of funding the fieldwork was already slowing down as this brief publication was being prepared. Soon after publication, fieldwork slowed even more and it was clear the initial target dates were not going to be met. By about 2005, a new threat appeared on the horizon: proposed new legislation about intellectual property rights and 'bio-prospecting'. Legal experts of the National Botanical Institute explained to the working group that continued fieldwork was almost certain to fall foul of the new legislation. The project ground to a halt.

Early in 2011 Ngwenya, Williams and I met a professional project co-ordinator and new funding sources were identified. As I write, these potential sources are being pursued and it is to be hoped that the ZBKP will be revived.

3 The structure of plant names

Introduction

This chapter is about morphology: it describes structure, not the structure of botanical entities (plants), but that of linguistic entities (words, names). In this introductory section we shall look at some confusion of terminology (both plants and words have roots and stems, for example) and also at the various parts of speech (nouns, verbs, adjectives, prepositions and so on) that can be incorporated into plant names.

From there we go on to look at the relationship between the underlying structure of a plant name and how this may manifest in a surface structure, the spoken or written word. Although all languages are considered primarily oral, in literate societies conventions of writing have become equally important and we need to consider some of the stylistic issues in the writing of plant names, such as whether to write a phrasal name as one word, as several words or as several words linked by hyphens.

The next section deals with structural complexities in vernacular names for plants and we shall look at English, Afrikaans and Zulu names, first those with single stems, and then those with phrasal stems.

All of this provides background for the main part of the chapter, a discussion of the structure of Zulu plant names. Here we shall start by looking at Zulu plant names within the context of the Zulu noun class system. In addition to the distribution of plant names within the various noun classes, we shall also consider how the noun class system helps to differentiate between plants and their fruits, and between individual plants and those growing together in a field, garden or grove.

The chapter then goes on to a detailed analysis of plant name stems in Zulu and will cover the following areas: simple stems; reduplicated simple stems;

complex stems; and compound stems. There is a special sub-section here to look at names of the **umthombothi** and **umsimbithi** type; that is, compounds incorporating the element -*thi* (tree).

Finally, following on from the section in the previous chapter on indigenous knowledge systems (IKS), which noted that in an oral system fixed forms do not necessarily exist, we shall look at 'free variation' in the morphology of Zulu plant names.

Terminology issues

As noted in Chapter 1, there are some areas of terminological overlap that can cause confusion when the disciplines of botany and linguistics are discussed simultaneously. The branch of linguistics that studies the structure of words (and all names are words) is called morphology, just as the study of the structure of plants is called morphology. Words have 'affixes', sub-divided into prefixes, suffixes and infixes, terms not applied to parts of plants. But words can also have roots and stems, and these are unquestionably terms used of plants. As this chapter is about the structure of plant names (rather than about the structure of plants), terms like 'morphology', 'root' and 'stem' are used in their linguistic sense unless very specifically stated otherwise.

All names are nouns

All names of whatever kind (personal, place, brand and plant) in whatever language are words from the word group known as 'nouns' in English. Names are not verbs, prepositions, adverbs or adjectives, although they may well <u>contain</u> these parts of speech in their underlying structure and, indeed, commonly do so. In such cases the name is usually a compound name or a phrasal name with a noun as its head (a 'noun phrase'). Although we shall look at this in far greater detail later, particularly when we come to Zulu plant names, let us consider briefly some examples from English, Afrikaans and Zulu that show various parts of speech used in plant names.

The following are names that consist simply of a noun:

From English: wisteria, cedar, clover, cypress, daffodil, elder, nettle, oak, pansy, pine and plum.

From Afrikaans: *laventel* (lavender), *kastaiing* (chestnut), *lelie* (lily), *tulp* (tulip) and *roos* (rose).

From Zulu: **umvumvu** (*Celtis africana*), **umdoni** (*Syzygium cordatum*), **umhlosinga** (*Acacia xanthophloea*), **isihlehle** (*Pachypodium saundersii*), **umdabu** (*Elephantorrhiza elephantina*) and **ingcino** (*Albuca setosa*).

The following names contain various parts of speech:

From English:
Blackjack [adjective + noun];
Jack-in-the-Pulpit (*Zantedeschia aethiopica*) [noun + preposition + article + noun];
Apple of Peru (*Nicandra physalodes*) [noun + preposition + noun (place name)];
Dainty White Wild Hibiscus (*Hibiscus meyeri*) [adjective + adjective + adjective + noun];
Little-Man-in-the-Boat (the inflorescence of *Strelitzia reginae*) [adjective + noun + preposition + article + noun].

From Afrikaans:
brandblaar (*Knowltonia vesicatoria*) [verb + noun];
kanniedood (*Celtis africana*) [negative verb + verb];
khakibos [adjective + noun];
klimopgras (*Olyra latifolia*) [verb + preposition + noun];
koning-van-kandië (*Haemanthus coccineus*) [noun + preposition + noun];
kruitjie-roer-my-nie (*Melianthus major*) [noun + verb + pronoun + negative particle].

Mixed English and Afrikaans:
Pride-of-De-Kaap (*Bauhinia galpinii*) [noun + preposition + article + noun].

From Zulu:
uhlambahloshane (what washes whitish; *Helichrysum adenocarpum*) [verb + adjective with diminutive suffix];
amafuthomhlaba (fat of the earth; *Callilepis laureola*) [noun + noun];
uphamaphuce (what gives and takes away; *Eclipta prostrata*) [verb + verb];
uvelabahleke (what appears and they laugh; *Lotononis corymbosa*) [verb + verb];
umhlabandlazi (what stabs the mousebird; *Aloe suprafoliata*) [verb + noun].

Underlying structure of plant names

Let us illustrate what is meant by underlying structures by taking two species of tree, *Dracaena aletriformis* and *Sclerocroton integerrimus*, and looking at the English and Afrikaans names of the first species, and the Zulu name of the second.

For *Dracaena aletriformis* Boon (2010: 58) gives Large-Leaf Dragon-Tree and *grootblaardrakeboom* as the English and Afrikaans names. '*Dracaena*' in the botanical name, 'dragon' in the English name and '*drake*' in the Afrikaans name are explained as derived from the Greek for a female dragon (*drakaina*) because the red resin of one of the *Dracaena* species is sold as 'dragon's blood'. Alternatively, it is named after Sir Francis Drake. The structures of the two vernacular names start with a head noun 'tree' (*boom*). This tree is qualified by the noun 'dragon' (*drake*), functioning as an adjective. The word 'leaf' (*blaar*) is likewise qualified by the adjective 'large' (*groot*). Finally, the noun-phrase 'large leaf' (*groot blaar*) qualifies the noun phrase 'dragon tree' (*drake boom*).

A tree diagram[1] of the syntactical relationship between the different elements in the underlying structures of Large-Leaf Dragon-Tree and *grootblaardrakeboom* would look like this:

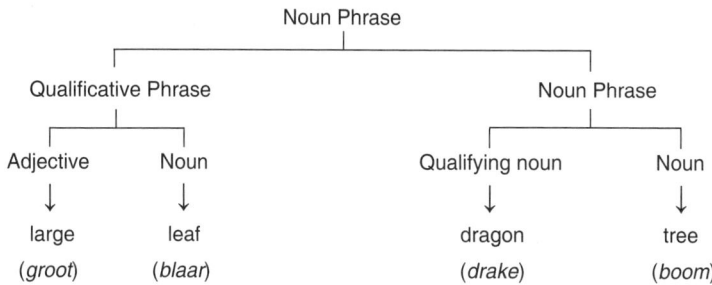

This structure manifests (realises or appears) as the single word *grootblaardrakeboom* in Afrikaans, with no indication of the relationship of the elements in the underlying structure. In English[2] it manifests as a four-word phrase in

---

1. Yet another linguistic term with botanical echoes!
2. In Boon (2010) at least. In the previous chapter we noted how individual authors and editors have to decide how to write vernacular names. Pooley has "Large-leaved Dragon Tree" (1993: 60); Van Wyk et al. have "large-leaved dragontree" (2011: 270); and Coates Palgrave (who refers to this tree as *Dracaena hookerana*) has "Large-leaved dragon tree" (1977: 86).

which the first two words are joined by a hyphen and the third and fourth words are also hyphenated, indicating at least the relationship between 'large' and 'leaf', and 'dragon' and 'tree'. It is up to the editor or author of a particular work to decide on where to use the upper case when it comes to vernacular names, although with botanical names it is universal convention to use upper case for the generic name and lower case for the specific name.

In the case of *Sclerocroton integerrimus* one of the two Zulu names Boon (2010: 254) gives is **umdlampunzi**, a name also used for *Shirakiopsis elliptica*. Doke and Vilakazi (1958: 153) give us a clue about the underlying structure of this name with their literal translation 'what the duiker eats'. The underlying structure of this name is *impunzi idla* [*izithelo zalesi sihlahla*] (the duiker eats [the fruits of this tree]). In other words, the name consists of the noun *impunzi* and the verb *dla*, with *impunzi* functioning as the subject of the verb. The object of the verb *dla* (eat) is the tree itself (*Sclerocroton integerrimus*) or at least its fruits. The normal word order of a Zulu sentence, where subject precedes verb, has been reversed in the manifestation of the name as **umdlampunzi**. Compare this name to the Zulu noun *umdlambila* from *dla* (eat) and *imbila* (rock rabbit). On the surface it would appear to have exactly the same structure as the tree name **umdlampunzi**, but this is not the case. According to Doke and Vilakazi this refers to a "species of venomous rock-cobra [or] species of hawk-eagle" (1958: 153). In this noun, *imbila* is the <u>object</u> of the verb *dla*, and the subject is the cobra and/or the hawk eagle.

There are a number of possibilities for the writing of this word:[3] **umdlampunzi, Umdlampunzi, uMdlampunzi, umDlampunzi, um-dlampunzi, um-Dlampunzi, Dlampunzi (um-)** are just some of the forms found in various sources that give Zulu plant names. The variations here are caused first by lack of consistency in the use of the upper case; second, the position of the capital if it is used; and third, treatment of the noun class prefix in Zulu. We note, though, that in all these variations, despite the underlying structure consisting of the two words *dla* and *impunzi*, the surface realisation is one word.

---

3. Without even taking into account font, font size, whether bold, italic, underlined or a combination of these, ink colour and so on, all of which are effectively used in Boon (2010).

The structure of Zulu plant names

Under this heading we shall look first at plant names as they occur in the Zulu noun class system and go on to to examine different types of noun stem structures.

Plant names in the Zulu noun class system

Zulu plant names, like all other nouns, belong to the 'noun class system' that is common to all languages belonging to the 'Bantu' group. Every noun consists of two basic parts:

- a class prefix indicating whether the noun is singular or plural and what concordial agreements will be used with other parts of speech; and
- the noun stem, which contains the meaning of the noun.

For example, in the pair *umuntu/abantu*, -*ntu* is the noun stem meaning 'human being', *umu*- is the singular class 1 prefix and *aba*- the class 2 plural prefix. *Umu*- and -*ntu* combine into *umuntu* with the meaning 'single human being' (i.e., a person), and *aba*- and -*ntu* combine into *abantu* with the meaning 'plural human being' (i.e., people). In the pair **umvumvu/imivumvu**, -*vumvu* is the noun stem meaning White Stinkwood (*Celtis africana*), *um*- the singular class 3 prefix, and *imi*- the plural class 4 prefix. **Umvumvu** thus means 'a single specimen of *Celtis africana*' or 'the species *Celtis africana*', and **imivumvu** means 'two or more *Celtis africana* trees'. In the pair **indulo/izindulo**, the stem -*dulo* means Bottlebrush Tree (*Greyia sutherlandii*), *in*- is the class 9 singular prefix, *izin*- the class 10 plural prefix. **Indulo** means one Bottlebrush Tree, or the species as a whole, and **izindulo** means two or more Bottlebrush trees.

The following set of contrasting sentences illustrates what is meant by 'concordial agreement':

b-onke l-aba ba-ntwana ba-mi aba-hle ba-khona[4]
'All these lovely children of mine are here'
y-onke l-e mi-vumvu ya-mi emi-hle i-khona
'All these lovely white stinkwoods of mine are here'

---

4. Standard Zulu orthography does not use these hyphens. They have been used here for ease of comparison.

z-onke l-ezi zi-ndulo za-mi ezin-hle zi-khona
'All these lovely bottlebrushes of mine are here'

Let us now look at the distribution of plants in the noun class system and then go on to examine how movement of names within the system can produce names for fruits on the one hand, and gardens, groves and fields on the other.

Distribution of plant names

Not all the Zulu noun classes are used for plant names. Classes 1 and 2 as in *umuntu/abantu* are used for human beings only. There are no plant names in these two classes. Class 1a as in *ubaba* (my father) with the plural in class 2a *obaba* (my fathers) is used for human beings only.

Class 3a[5] as in **ubathini**, one of several names for the Pigeonwood, *Trema orientalis*, has the same plural as class 1a, thus **obathini** (two or more Pigeonwood trees). Class 3, with its plural in class 4, unquestionably has the greatest number of plant names. Louwrens (1999: 299) claims that the 'meaning' of classes 3 and 4 (if a noun class can be said to have a meaning) is 'growing things'. Hundreds upon hundreds of examples can be given, but let us just note one example, that of **umdoni** (Waterberry, *Syzygium cordatum*) with the plural **imidoni**.

Class 5, with its plural in class 6, has quite a number of plant names such as **ihlukwe** (White Arum Lily, *Zantedeschia aethiopica*) with the plural **amahlukwe**. Occasionally, class 6, although a plural class, is used for the species or a single specimen as in **amasethole** (literally, sour milk of the heifer, Coast Red Milkwood, *Mimusops caffra*).

Class 7, with its plural in class 8, also has number of plant names such as **isidikili**, one of several names for the Lesser Yellow Head (*Gnidia kraussiana*), with the plural **izidikili**.

Class 9, with its plural in class 10, is also used for plant names. An example is **intelezi**, a word used for numbers of plants used for 'sprinkling' traditional protective charms. The plural is **izintelezi**.

Class 11, which shares class 10 as a plural, has few plant names. Broom Asparagus (*Asparagus virgatus*) goes by the Zulu name **unwele**, with the plural **izinwele**.

---

5. Not all Zulu-language scholars and Zulu grammars accept a distinction between classes 1a and 3a.

Classes 12 and 13, used to form diminutives in the Bantu languages spoken north of the Limpopo River, have long been missing from the Bantu languages of South Africa.

Class 14 is usually used to form abstract nouns, such as *ubuntu* 'humanity' (cf. *umuntu/abantu*), *ubude* (length/height < *-de* (long, tall)) and *ubuhlungu* (pain < *-hlungu* (painful)). This noun class contains very few plant names. An example is **ububendle** (Doll's Protea, **Dicoma zeyheri*). There is no equivalent plural for this class.

As in all Bantu languages, class 15 in Zulu is used exclusively for the infinitive forms of verbs. The normal prefix form is *uku-* as seen in *ukufa* (to die, dying and death), *ukudla* (to eat, eating and food) and *ukubona* (to see, seeing and vision). There are no plant names in this class.

Plant name stems occurring in more than one noun class

It can happen that the same noun stem can be found in more than one noun class as in **intingwe** (class 9) and **untingwe** (class 3a). When this happens, either the reference of the name stays the same (i.e., the different versions refer to the same plant), or the variations will have different references (i.e., the different versions refer to different species). In the case of **intingwe** and **untingwe**, both names refer to *Anemone caffra*.

The reference stays the same:
- **ingonswane** (class 9), **isigonswane** (class 7) and **umgonswane** (class 3) all refer to various species of Wild Fig-Tree (Doke and Vilakazi 1958: 258). In the case of **umgonswane**, Doke and Vilakazi specify "rock-splitting" fig trees such as **Ficus sonderi* and *Ficus ingens*;
- **inhlakahla** (class 9) and **u(lu)hlakahla** (class 11) both refer to **Agapanthus umbellatus*;
- **umhlonhlwane** (class 3) and **inhlonhlwane** (class 9) both refer to "several species of small *Euphorbia* trees" (334).

The reference changes:
- **ingceba** (class 9) is a Wild Banana plant (**Strelitzia augusta*), but **igceba** (class 5) is a species of Marsh Rush used for mat making;
- **umhlezane** (class 3) is a species of bush, *Ochna arborea*, **uhlezane (o-)** (class 1a) is Hickory-King Mealie and **inhlezane** (class 9) a species of long succulent grass chewed by children (Doke and Vilakazi (1958: 328).

A curious mixture can be seen in two derivatives from **imfika**, a species of *Hermannia* plant. **Imfikane** (class 9) refers first to "small species of *Hermannia*" and second to a "species of grass", while **umfikane** (class 3) refers to "various species of grass, e.g. **Andropogon sorghum*" (Doke and Vilakazi 1958: 206).

Occasionally the reference changes from a botanical referent to a zoological one: **igekle** (class 5) is *Crassula acinacifolia*,[6] the stem of which is used to make hemp pipes and flutes but *ingekle* (class 9) is a white heron or egret; while **umhlangothi** (class 3) and **unhlangothi** (class 3a) both refer to a "species of forest tree, *Protorhus longifolius* [= *longifolia*]" (Doke and Vilakazi 1958: 320), while *inhlangothi* (class 9) is a species of shark.

Names of fruit

It frequently happens that the name of a tree is in one noun class (usually class 3, with the prefix *um-*), while the same noun stem, when moved to a different noun class, refers to the fruit of that tree. For example, Doke and Vilakazi give **umhlalanyathi** as "species of large shady tree, *Grewia occidentalis*" and **ihlalanyathi** as "fruit of the *Grewia occidentalis*" (1958: 315). Further examples are:

- **umhlakuva**: Castor-Oil Bush (*Ricinus communis*), **uhlakuva/inhlakuva** its fruit;
- **umkhiwane**: Fig Tree (*Ficus sur*), **ikhiwane** its fruit (see Plates 13 and 14);
- **umgwenya**: (*Harpephyllum caffrum*), **igwenya** its fruit;
- **umdoni**: Water Myrtle (*Syzygium cordatum*), **indoni** its "black, edible berries";
- **umgulukunqa**: **Strychnos gerrardii* (Monkey-Orange Tree), **igulukunqa** its fruit;
- **umhlala**: *Strychnos spinosa* (Monkey-Orange Tree), **ihlala** its fruit.

Occasionally the same noun stem is found in three different noun classes, with extra subtleties of meaning as in the case of the Maroela Tree, where Doke and Vilakazi give **umganu** for Maroela (*Sclerocarya birrea*), **iganu** the fruit of the Maroela Tree and **amaganu**, beer made from ripe *umganu* fruit.

---

6. This is the name given by Doke and Vilakazi. Beaumont has not been able to trace it and wonders whether Doke and Vilakazi meant *Crassula acinaciformis* or *Crassula acutifolia*.

The noun stem -*thunduluka* also occurs in three different noun classes: class 3 **umthunduluka**, the fruit tree *Ximenia caffra*; class 5 **ithunduluka**, which refers to the fruit of *Ximenia caffra*; and class 9 **intunduluka**, a stone in the fruit of *Ximenia caffra*.

In all likelihood, most tree names refer to the tree or plant as a whole, including all its parts such as bark, flowers, roots, seeds and fruit. As we have seen above, occasionally Doke and Vilakazi offer a noun stem in two different classes, indicating separate words for fruit and tree. In one rare example they give a single word for a tree and stress that this refers to both the whole and the part: "**ibhonsi**: fruit and plant of *Salacia alternifolia*" (1958: 45).

Zulu plant names with related words for fields, gardens, thickets and groves

A different kind of meaning change takes place when a name for a plant is moved from its 'natural' noun class into noun class 7 with the prefix *isi*-. For example, let us look at the **idumbe** tuber (*Colocasia esculenta*), known to most English-speaking residents of KwaZulu-Natal as the madoombie, an adaptation of the plural form **amadumbe**. When the class 5 noun **idumbe** moves into class 7 as **isidumbe**, it refers to a garden of **idumbe** tubers. The word for sugarcane is **umoba** while **isimoba** refers to a field of sugarcane.

Gerstner is one of the few writers (and this includes writers of Zulu grammars) to note the phenomenon:

> Remember that the prefix *isi*- is often used to express that there is a clump or grove of trees, e.g. *isiNga* instead of *umuNga*. This pure grammatical form has here been much neglected (1938: 215).

As with the plant-fruit pairs above, the potential shift of a plant name noun stem into class 7 to indicate a field or garden of such plants is by no means a predictable and regular occurrence:

- **umkhoba**: Yellowwood Tree (*Podocarpus latifolius*) while **isikhoba** is a forest of Yellowwood trees;
- **imfe**: Sweet-Reed (**Andropogon sorghum*) while **isife** is a plot of Sweet-Reed;
- **ubhatata**: Sweet Potato while **isibhatata** is a field of Sweet Potatoes;
- **ingceba**: Wild Banana Tree (**Strelitzia augusta*), but **isigceba** is first a Wild Banana Tree (**Strelitzia augusta*) and second a plantation of **S. augusta* trees.

In the following two examples we see the 'default' form of the plant name moved to <u>two</u> other noun classes, giving both the fruit and the field forms:

- **amaqathe**: shrub *Pachystigma pygmaea* [= *pygmaeum*] and **iqathe** the fruit of *P. pygmaea* with **isiqathe** a place overgrown with *P. pygmaea* plants;
- **umhlalane**: sp. of bush, *Strychnos gerrardii*, **ihlalane** its fruit and **isihlalane** a thicket.

Change of noun class means change of species

Occasionally, moving from one noun class to another means a change in species as well. For example, **imbuzana** is a class 9 noun meaning *Ipomoea albivenia*. When the same stem *-buzana* is moved to class 7 as **isibuzana** it means species of *Andropogon* grass. Another example involves the noun stem *-cakathi*. When this moves into class 5 as **icakathi**, it refers to a "species of plant, *Agapanthus umbellatus*" (Doke and Vilakazi 1958: 99). But when the same stem goes into class 7 as **isicakathi** it refers to a "species of plant, *Salvia scabra*, used as a purgative" (99). It is worthwhile noting here that when the same stem is moved into class 3 as **umcakathi**, this is used as a *hlonipha* term for *umuthi* (tree, medicine).[7]

Different types of Zulu noun stems

Under this heading we shall look at the following types of noun stems: (1) simple stems; (2) reduplicated simple stems; (3) complex stems; (4) phrasal or compound stems; and (5) phrasal stems incorporating *-thi*.

Simple stems

These cannot be sub-divided into meaningful parts. For example, the stem *-bhucu* in the plant name **ibhucu** (*Bulbine abyssinica*) cannot be broken into the meaningful units *bhu* and *cu*. Similarly, the stem *-jobo* in the plant name

---

7. The concept of *hlonipha* language probably deserves far more than footnote discussion. Zulu women are obliged by social rules of respect to avoid using words that sound the same as the names of their close male relatives. To overcome these linguistic strictures they have evolved a separate vocabulary for many common Zulu words, known collectively as '*hlonipha* terms'.

injobo (*Cyrtanthus breviflorus*) does not consist of the meaningful units *jo* and *bo* as *-jobo* is a single indivisible unit.

All the following plant names from Pooley (1998) have simple stems:
- **inhlaba** *Aloe chabaudii* (Chabaud's Aloe)
- **icena** various species of aloe
- **uqonsi** *Eriosema salignum* (Narrow-Leaved Eriosema)
- **ibhonsi** *Salacia kraussii*
- **ugweje** and **umdabu** *Elephantorrhiza elephantina* (Elephant's Root)
- **ishwaqa** *Pelargonium alchemilloides* (Pink Trailing Pelargonium)
- **isiphunga** *Lotus discolor* (Coral Plant)
- **insololo** *Sphenostylis angustifolia* (Wild Sweetpea Bush)
- **isidenda** and **udoye** *Maesa alnifolia* (Dwarf Maesa)
- **isikhonde** and **isiphofu** *Asclepias multicaulis* (Doily Cartwheel)
- **isinama** *Priva cordifolia*
- **inkomfe** two species of *Hypoxis*: *H. multiceps* (Winter Star-Flower) and *H. rigidula* (Silver-Leaved Star-Flower)

Some tree names from Boon (2010) with simple stems are:
- **umhluhluwe** *Dalbergia armata* (Thorny-Rope Flat-Bean)
- **umsinsi** *Erythrina caffra* (Coast Coral-Tree)
- **umvumvu** *Celtis africana* (White Stinkwood)
- **umunga** various species of *Acacia*
- **ugagane** *Dichrostachys cinerea* (Sickle Bush)
- **umhlwakela** *Drypetes arguta* (Water Ironplum)
- **umhlonhlo** various species of *Euphorbia*
- **umhlepha** *Sclerocroton integerrimus* (Duiker-Berry)

Plant names with simple stems do not usually have any deeper or underlying meaning.[8] The meaning of **umvumvu** is 'the tree species *Celtis africana*' or 'a single specimen of the species *Celtis africana*'. Compare this to, say, the tree **umphindamshaya**. The primary meaning here is 'the tree species *Adenia gummifera*' or 'a single specimen of the species *Adenia gummifera*', but there is a secondary, underlying meaning: 'what hits him again', a reference to the use of parts of this tree in counter-spells against witchcraft.

---

8. Layers of meaning are discussed in Chapter 4.

There are, however, a few simple stemmed nouns that have other, primary, meanings relating to the plant. For example, the Zulu name of the Common Buttercup (*Ranunculus multifidus*) is **uxhaphozi**. This is a name with a single stem, but as the plant is found "in damp ground near streams and marshes" (Pooley 1998: 252) clearly the name is linked to the Zulu word *ixhaphozi* (marsh or swamp). Similarly, the Zulu plant name **unwele**, which Pooley gives for both *Sutherlandia montana* and *Asparagus virgatus* (1998: 58, 104), is clearly linked to the meaning of the common noun *unwele* (single strand of hair). Pooley points out that the leaves of *Asparagus virgatus* are "threadlike".

Plant names with reduplicated stems

There are a number of plant names in which the stem appears to have been duplicated, as in **isitokotoko** (*Sansevieria hyacinthoides*) (Pooley 1998: 100). Reduplication is a common structural feature of Zulu words and is sometimes meaningful and sometimes not. When a verb root is duplicated, this has the effect of lessening the action of a verb: *phuzaphuza* (to sip at a drink, cf. *phuza*, drink); *hambahamba* (stroll along, cf. *hamba*, walk); *bonabona* (glimpse, cf. *bona*, see). When a plural noun stem is duplicated, however, this augments the plural: *izintaba* (mountains or hills, cf. *izintabantaba*, one mountain after another or very hilly countryside); *imifula* (rivers, cf. *imifulamfula*, river delta with the main river broken up into a number of smaller channels); *imisindo* (sounds, cf. *imisindomsindo*, babble of dozens of different sounds).

Many nouns have reduplicated stems, but with no such indication of multiplicity. Such nouns are usually derived from an ideophone or other part of speech: *isiphalaphala* (woman with lovely eyes < *phála*, of roaming eyes); *isiphoshophosho* (garrulous person or gossip < *phósho*, of gossiping, chattering continually); *impoqompoqo* (anything brittle, easily snapped < *phóqo*, of snapping, breaking). The etymology is not always clear: for example, *isigomegome* (person who stands firmly by his word). Doke and Vilakazi don't say so, but this is probably linked to the verb *goma* (take an oath, swear). Another instance is *isigonogono* (earwax, with no obvious etymology).

Plant names with reduplicated stems are very common and for some of them some sort of underlying meaning can be detected, or at least suggested:

- **impishimpishi**: *Aspalathus chortophila* (Tea Bush). Doke and Vilakazi (1958: 666) give five meanings for the ideophone *phíshi*, but none seems to have any meaning that could connect to a plant;

- **umozamoza**: *Pavonia leptocalyx*. Doke and Vilakazi give **umozomozo**, linking it to the obsolete ideophone *mózo* (of smiling), almost certainly linked to the use of this plant as a love charm;

- **intikintiki**: *Cyphia longifolia*. Doke and Vilakazi give "jellylike, congealed substance < *thíki* 'of quivering'" (1958: 794). Can this be related to the plant?

- **iklabeklabe**: *Lactuca inermis* (Wild Lettuce). Doke and Vilakazi (1958: 429) give this as the name of four different species and say the word is derived from the ideophone *klábe* (of stealthy glance, of cutting in slices, of mocking with closed hands). Which of these meanings could be related to a plant? Perhaps slicing up the lettuce for a salad?

- **insulansula**: *Eriospermum mackenii* (Yellow Fluffy-Seed). Doke and Vilakazi say "Protective charm against witchcraft, charm for blinding, distracting attention, made from herb *Spermacoce natalensis*" (1958: 769) and derived from the ideophone *súla* (of blinding, distracting attention);[9]

- **isihlekehleke**: *Euphorbia clavarioides* (Lion's Spoor). Doke and Vilakazi give this word only as "anything with a wide-open mouth (< *hléke*, of being wide open)" (1958: 326);

- **ibongabonga**: *Solanum mauritianum* (Bugweed), which I know as **impongompongo**. The Xhosa word is *umbangabanga*. It is unlikely that the name is related to the verb *bonga* (give thanks, be grateful, utter praise poetry). It is equally unlikely that the plant name is related to the noun *impongo* (goat-ram);

- **isihomohomo**: *Philenoptera violacea* (Apple-Leaf). Doke and Vilakazi record this word as the name of a "species of large tree found in Swaziland" (1958: 344), but make no suggestions as to possible derivation;

- **isishwashwa**: *Xylotheca kraussiana* (African Dogrose). Doke and Vilakazi recognise this as the name of the Cape Dog-Rose, but again make no suggestions about derivation;

- **umhlungahlunga**: *Vernonia tigna* (Mountain Vernonia). Doke and Vilakazi give both **u(lu)hlangahlunga** and **u(lu)hlunguhlungu** for the "shrub *Vernonia corimbosa*" (1958: 338, 339) and as with the last two examples offer no ideas about derivation. Is there any way this shrub

---

9. Very probably also related to the verb *sula* (to erase).

could be related to the "sugar-bush, *Protea hirta*" for which Doke and Vilakazi (1958: 338) give the Zulu name **isihlunga**?

Complex stems

Complex stems <u>can</u> be sub-divided into smaller meaningful units. For example, the stem -*miselo* in **isimiselo** (*Gloriosa superba*) can be broken down into *m*[*a*] (stand) + *is* (cause to) + *el* (on behalf of) + *o* deverbative noun marker, that is 'the thing that causes something to stand up on someone's behalf' (in this case a plant used medicinally to treat impotence and so cause erections where they had been lacking before). The stem -*thangazane* in the plant name **uthangazane** (*Cucumis hirsutus* or Wild Cucumber) can be sub-divided into *thanga* (pumpkin) + *azi* (biggish) + *ane* (smallish), that is a plant resembling a medium-sized pumpkin. The same two suffixes, -*kazi* and -*ane*[10] occur in the name **umthongakazane** (*Pyrenacantha scandens*). In another name for this plant, **umkhokhothwane**, only the suffix -*ane* has been added. *Pyrenacantha scandens* has two more Zulu names and these are also complex, but in a different way: **umnakile** is formed from the verb root *naka* (be aware of) with the perfect tense suffix -*ile*, while **umsekelo** is derived from the verb *sekela* (support) with the noun-forming suffix -*o*. Going back to plant names with the combined suffixes -*azi* and -*ane* we note both **ishashakazane** and **isijojokazane** as names for the Common Buttercup (*Ranunculus multifidus*).

USE OF -*MA*-

The name **umangwazane** (*Gomphocarpus physocarpus*) also has the suffix -*ane*, but the name-forming prefix -*ma*- as well. This prefix has a number of different meanings in the Zulu language:

- It could be the short form of the class 6 noun class prefix *ama*-, and this it what it is in the names **umathunga** (used for the two *Kniphofia* species *K. linearifolia* and *K. porphyrantha*), which is derived from *amathunga* (wooden milk pails),[11] **umahogwe** (*Kalanchoe crenata*,

---

10. -*azi* and -*kazi* are variations of the same suffix, as are -*ane* and -*ana* of a different suffix. Such variations, which cause no change in meaning, are known as allomorphs.

11. Zukulu et al. (2012: 44), who give the Pondo name *umathunga* for *Eucomis autumnalis*, say that a decoction of the bulb is used to treat broken limbs both human and animal, hence the name meaning 'sewing together'. They derive the plant's name from the verb *thunga* (sew).

Yellow Hairy Kalanchoe) derived from *amahogwe* (bitter wild lettuce plants), **umadolwane** (*Plectranthus laxiflorus*, Citronella Spur-Flower) from *amadolo* (knees)[12] and **umadlozana** (*Argyrolobium tomentosum*, Velvety Yellow Bush-Pea), which is derived from *amadlozi* (ancestral spirits). These last two names both have the suffix *-ana/-ane*;

- It could be the verbal prefix *ma-* used in exhortations, as in *masihambe* (let us go) and this is certainly the case for the plant name **umakuphole** (*Pentanisia prunelloides*, Broad-Leaved Pentanisia). Here the verb is *phola* (cool down) and the underlying meaning of the word (let it cool down) is a reference to the use of this plant as 'cooling medicine'.[13] In the plant name **umayime** (*Osyridicarpos schimperianus*) *-ma-* is used in the same way. The underlying meaning 'let it stop' also refers to medicinal use. This plant is also known as **umalala** and this suggests yet another meaning of the prefix *-ma-*;

- It could be simply a name-forming prefix with no clear meaning as in the descriptive compound *umahlekehlathini* (he who laughs in the bush) given to a heavily bearded person, and in the equally descriptive *umalalepayipini* (he who sleeps in the culverts), a word indicating a homeless tramp. This is the same prefix used with the maiden clan names of married women, used commonly as a form of address. In this form of address a woman from, say, the Ntuli clan will become known as uMantuli wherever she might marry.[14] Plant names using *-ma-* in this sense include the example given above, **umalala**, derived from the verb *lala* (lie down, sleep); **umathinta** (**Eriospermum abyssinicum*) based on the verb *thinta* (touch); and **umanqanda** (*Asclepias gibba*, Humped Turret-Flower), derived from the verb *nqanda* (prevent, turn away).

USE OF *-NO-*

The prefix *-no-* is frequently used to make names of all kinds. When added to a noun to form a personal name, this always (in today's Zulu society at least) forms a female name, as in uNobuhle (< *ubuhle*, beauty), uNonhlanhlana (< *inhanhla*, luck), uNomfundo (< *imfundo*, education) and uNomvula (< *imvula*, rain). When forming the names of animals, birds and insects

---

12. From the shape of the flower according to Beaumont (personal communication).
13. 'Cooling medicines' are described in some detail in Chapter 7.
14. See Koopman (2002: 23) for further details.

there is no female gender specificity: *unogwaja* (rabbit < *gwaja*, dart about), *unogandilanga* (tinker barbet < *ganda*, pound + *ilanga*, [all] day [long]) and *unomponjwana* (species of horned beetle < *i[zi]mpondo*, horns + *-ana*, little).

Examples of plant names that use this prefix are first **unompingi** (*Cyrtanthus brachyscyphus*, Dobo Lily, cf. **impingizana** *Cyrtanthus contractus*, Fire Lily). The meaning of the probable root *impingi* cannot be traced. Second, **unobebe** (*Eugenia albanensis*, Dwarf Grassland Eugenia) is based on the noun *ibebe* (pleasant food).

The prefix *-no-* is used in a most intriguing relationship with two Zulu ideophones relating to the way food is eaten. Doke and Vilakazi (1958: 430, 432) give the ideophone *klámu* (of biting crisp fruit or vegetable) and the ideophone *kléshe* (of biting fatty meat). These two ideophones are combined with *-no-* to form the plant names **unoklamu** and **unokleshe**. Curiously, the two ideophones also combine with each other, to form the name **uklamkleshe**. Pooley's (1998) index shows that all three names are used for a number of plants. Some of them, as the list below shows, use all three names. The numbers after each botanical name indicate a page in Pooley (1998):

Disa polygonoides (48)	**uklamkleshe**		
Disa chrysostachya (48)	**uklamkleshe**		
Habenaria epipactidea (114)	**uklamkleshe**	**unokleshe**	
Habenaria falcicornis (116)	**uklamkleshe**		
Satyrium sphaerocarpum (118)	**uklamkleshe**	**unokleshe**	**unoklamu**
Satyrium longicauda (362)	**uklamkleshe**	**unokleshe**	**unoklamu**
Satyrium macrophyllum (364)	**uklamkleshe**	**unokleshe**	**unoklamu**

Given the 'biting references' in the two ideophones that lie at the base of these three plant names, it is interesting to note Doke and Vilakazi's entry for **umklamkleshe**:

> **umklamkleshe** (also **uklamkleshe**, and with *no-* to form **unoklamkleshe**): Species of ground orchid; edible types: *Habenaria foliosa, H. caffra, Satyrium longicauda, S. macrophyllum, S. sphaerocarpum*; non-edible types, *Disa chrysostachys, D. polygonoides*. [cf. *unoklamu, unokleshe*; v.l. *uklamkleshe*] (1958: 430).

Not knowing the plants referred to, I can only assume that the 'edible types' were first assigned the 'crispy-fatty-munchy' epithets, and the 'non-edible types' were later assigned the same names on the grounds of similar appearance (as opposed to similar munchiness). Incidentally, Doke and Vilakazi's entries for **unoklamkleshe** and **unokleshe** are the only indications in the literature that these plants have edible parts: "species of ground orchid with edible tubers" (1958: 583).

We end this section on the use of -*no*- with a reminder of the name **umfana kaNozihlanjana** (*Stylochiton natalensis*, mentioned in Chapter 2). Doke and Vilakazi enter this plant under **um' fana-kaHlanjana** and say, "also heard as **umfana-kaNozihlanjana** and **umfana-kaSihlanjana**" (1958: 199). As I pointed out earlier, there are so many free variables in this name that there are at least 48 possible spellings, all potentially correct. The relevant variation here is of course the presence or absence of the name formative -*no*-.

VARIOUS OTHER VERBAL AND NOMINAL AFFIXES

A variety of verbal and nominal affixes (that is, both prefixes and suffixes) can be seen in the following Zulu plant names with complex stems. The name **umbonisela** refers to *Chamaecrista mimosoides* (Fishbone Dwarf Cassia) and the stem has the structure [*m* (him) + *bon*[*a*] (see) + *is* (cause to) + *el* (do on behalf of) + *a* (the final -*a* of the verb]. The name thus has the underlying meaning 'that which causes someone to see on behalf of others'. The verbal 'causative extension' -*is*- is also found in the two plant names **igqokisi** (*Indigofera hilaris*, Red Indigo Bush) used with the verb *gqoka* (get dressed) to give the meaning 'that which causes someone to get dressed'; and **isiwisa** (*Tephrosia elongata*, Orange Tephrosia) from the verb *wa* (fall over, fall down) giving the meaning 'that which causes someone to fall', a name likely to be related to its use in witchcraft. The plant is also used in concoctions designed to cause or prevent the fall of hail.

Compound stems

A 'compound noun' can be defined as a noun that has two or more parts, each of which can function on its own in a sentence. In the English word 'milking', 'milk' can stand on its own, but '-ing' cannot, so 'milking' is not a compound. On the other hand, in the word 'milkman', both 'milk' and 'man' can stand on their own, so 'milkman' is a compound. In the Zulu plant name **umbonisela** (see above), we can divide the word into *um* + *bon*[*a*] + *is* + *el* + i, but only the verb *bon*[*a*] can stand on its own. It is not a compound. However, in the plant name **usingalwesalukazi** (*Gomphocarpus physocarpus*) both *usinga* (thread) and *lwesalukazi* (of the old woman) can stand alone, so this word is a compound.

Compound plant names in Zulu use many different parts of speech in a variety of combinations. The following are just a few of the possibilities. Many of these examples have already been given above under various headings and some will be given later in this book. They are: noun + noun; noun + verb; noun + adjective; noun + possessive; verb + noun; verb + adverb; and verb + verb. Examples are:

- **ubuhlungubendlovu**: *Strophanthus petersianus* (Sand Forest Poison Rope) [*ubuhlungu* (pain) + *ba* (of) + *indlovu* (elephant) (the *-a* of *ba-* and the *i-* of *indlovu* coalesce to form *-e-*)];
- **usingalwesalukazi**: *Gomphocarpus physocarpus* (Milkweed, Balloon Cottonbush, Hairy Balls; also **umangwazane**) [*usinga* (thread) + *lwa* (of) + *isalukazi* (old woman)];
- **amafuthomhlaba**: *Callilepis laureola* (Ox-Eye Daisy) [*amafutha* (fat) + *a* (of) + *umhlaba* (the earth)];
- **umhlabangubo**: *Bidens pilosa* (Common Blackjack) [*um* + *hlaba* (stab) + [*i*]*ngubo* (blanket, clothing)];[15]
- **amabelejongosi**: *Polystachya sandersonii* [*amabele* (breasts) + *a* (of) + *ijongosi* (young maiden)];
- **imfeyamasele**: *Eulophia parviflora* [*imfe* (Sweet-Reed) + *ya* (of) + *amasele* (frogs)];
- **indabuluvalo**: *Kalanchoe paniculata* (Large Orange Kalanchoe) [*in* + *dabul*[*a*] (cause) + *uvalo* (fear)];
- **umlomomnandi**: *Argyrolobium tomentosum* (Velvety Yellow Bush-Pea) [*umlomo* (mouth) + *omnandi* (which is pleasant)];
- the Heart-Leaved Eriosema (*Eriosema cordatum* and other *Eriosema* spp.) that has two Zulu names, both of which are compounds: **ubangalala** [*u* + *banga* (cause) + *lala* (sleep)]; and **umhlabankunzi** [*um* + *hlaba* (prick) + *inkunzi* (bull)]. Both names refer to the aphrodisiacal qualities of this plant;
- the Small Yellow Gerbera (*Gerbera piloselloides*) with three Zulu names, each of which is a compound: **indlebeyempithi** [*indlebe* (ear) + *ya* (of) + *impithi* (duiker)]; **uhlangolumpofu** [*uhlanga* (reed) + *olumpofu* (which is pale)]; **umoyawezwe** [*umoya* (air, wind) + *wa* (of) + *izwe* (country)].

---

15. The fruits have barbs that attach to clothing, a method of seed dispersal.

Phrasal names

The examples of Zulu plant names with compound stems given immediately above could all be regarded as phrasal names in the sense that their underlying structures consist of two or more words in a phrase, even though in writing these elements form a single word.

I would like now, though, to use the term phrasal names in a slightly different sense to refer to the situation in Zulu folk-taxonomy where a plant name (which, as we have just noted, can itself be a compound noun and so phrasal) is distinguished by various epithets. Such names usually consist of two words (occasionally three), in which the first refers to a broad group of plants, perceived to have some similarity (not necessarily in strict botanical terms), and the second modifies the first in terms of colour, shape, size, habitat or other qualification so as to distinguish between the closely related species in a genus or between plants that are superficially similar. For example, Pooley (1998: 472, 436) gives **uhlalwane** as a name for *Pycnostachys reticulata* and **uhlalwane oluncane** (little uhlalwane) as a name for *Peristrophe cernua*. The following are examples:

Size distinctions:

- **umusa** *Stachys nigricans*, **umusa omkhulu** (big umusa) *Buchnera simplex* and **umusa omncane** (little umusa) **Adhatoda Andromeda*;
- **impimpi** *Acrolophia cochlearis*, **impimpi encane** (little impimpi) **Oeceoclades mackenii* and **impimpi enkulu** (big impimpi) *Satyrium parviflorum*.

Colour distinctions:

- **umahesaka-obomvu** (red umhesaka) *Thesium pallidum* ("flowers small, creamy white . . . whole plant has a yellowish appearance at certain times of the year") (Pooley 1998: 250), **umahesaka-omhlophe** (white umahesaka) *Agathosma ovata* ("flowers white to lilac") (402), **umahesaka-onsundu** (brown umahesaka) *Muraltia lancifolia* (Purple Heath: "Flowers pink") (406)[16]
- **indola** *Gnidia calocephala* ("Flowers white, hairy") (Pooley 1998: 158), **indola ebomvu** (red indola) *Hibiscus pedunculatus* ("Flowers

---

16. As can be seen from the description of the flowers given by Pooley, there seems no reason for the colour distinctions indicated by the Zulu names.

pale to deep pink or lilac") (408), **indola empofu** (pale indola) *Pavonia burchellii* ("flowers white to orange") (284).

Habitat distinctions:

- **ihlamvu** *Gloriosa superba, Callilepis laureola* and others, **ihlamvu lasenhla** (down-country ihlamvu) *Sandersonia aurantiaca*, **ihlamvu lasolwandle** (coastal ihlamvu) *Gloriosa superba*, **ihlamvu lehlathi** (forest ihlamvu) *Littonia modesta* and **ihlamvu elimpofu lasenkangala** (pale open-plain ihlamvu) *Disa stachyoides*.

A single plant name is often used for a variety of different species in different genuses because of some perceived similarity. When further qualified, however, the new phrasal name will normally only refer to one species. For example, the name **ishongwe** is used for the following unrelated plants: *Asclepias cucullata, Xysmalobium undulatum, Pachycarpus coronarius, Pachycarpus dealbatus* and *Pachycarpus concolor*. When further qualified, though, only one species is referred to: **ishongwe elibomvu** (red ishongwe) *Pachycarpus asperifolius* Large Red Milkwort, also **ishongwe elincane** (little ishongwe);[17] **ishongwe elibomvu elikhulu** (big red ishongwe) *Pachycarpus natalensis*; and **ishongwe elincane elibomvu** (little red ishongwe) *Schizoglossum atropurpureum*.

Phrasal names usually only have two words, although as can be seen from the examples above some have three words. Further examples are **umnduze wotshani obomvu** (red grass lily) *Disa chrysostachya*, **impingizane encane empofu** (little pale impingizane) *Cyrtanthus mackenii* (Ifafa Lily) and **ilabatheka elikhulu elibomvu** (big red ilabatheka) *Satyrium parviflorum*.

Compounds with -thi

Van Warmelo comments on compound tree names that incorporate a vernacular name for tree:

> There is another species of hardwood which the Venda call *mu-simbi-ri* "iron tree", from *-simbi* "iron" and *-ri* "tree". In Tsonga there is *nsimbi-tsi* "Lebombo Ironwood" (*Androstachys johnsonii*), but the *-tsi* instead of *-rhi* points to Swazi origin or influence (1976: 93).

---

17. Why the little ishongwe should have the English name Large Red Milkwort, I cannot say.

Using the metaphor of iron in the names of trees with very hard wood seems to be a widespread practice and in all examples from various languages a word for iron is linked to a word for wood. The Zulu name for *Millettia grandis* is **umsimbithi** with the structure *um-* + [*in*]*simbi* (iron) + [*umu*]*thi* (tree), exactly as in the Venda and Tsonga examples in the Van Warmelo quotation earlier. Boon (2010: 160) and Coates Palgrave (1977: 310) both confirm that the wood is very hard and heavy. Coates Palgrave (2002: 415) says of the Lebombo Ironwood (*Androstachys johnsonii*) that its wood is extremely hard and that its 'Rhodesian' name is Simbi Tree. Wild (1952: 54) gives, for the same tree, the Ndau[18] names *Sinbeti* [*sic*], *muZembiti* and *muZibiti*, all clearly based on the same [iron + tree] structure.

English common names for trees using the iron metaphor include Rhodesian Ironwood (*Colophospermum mopane*)[19] and Ironwood (*Olea capensis*), which has the Afrikaans name *ysterhout* (iron wood) (Boon 2010: 476). Boon (2010: 454) gives the scientific name of the White Milkwood as *Sideroxylon inerme*, the genus name being a compound of the Greek *sideros* (iron) + *oxylon* (wood).

The Zulu word **umsimbithi** mentioned earlier is just one of several Zulu names for plants incorporating the noun *umuthi*. This word is usually taken to mean 'tree' or 'medicine', but as Van Warmelo points out after giving the examples *indlolothi* (species of bulb), *ingulathi* (species of lily) and *inhlolothi* (species of grass), "the three last-named remind us that *umuthi* means not only tree, but plant in its wider sense" (1976: 94).

The following are all Zulu names for trees:

- **umkhovothi**: Thorny Elm, *Chaetachme aristata*;
- **uqhambathi**: *Protea roupelliae*;
- **isifithi**: Forest Camwood, *Baphia racemosa*;
- **umgodithi**: Bead-Bean, *Maerua angolensis*;
- **umfomothi**: Lebombo Wattle, *Newtonia hildebrandtii*;
- **ihlunguthi**: Copper-Stem Corkwood, *Commiphora harveyi*, cf. the Pedi name *mu-hloko-re* (bitter tree) for *Clerodendrum glabrum* (Van Warmelo 1976: 93);
- **uthovothi**: Forest False-Nettle, *Acalypha glabrata* also called **umthombothi**, of which more later;

---

18. Ndau is one of the main dialects of the Shona language spoken in Zimbabwe.
19. Wild (1952: 32). Curiously, later we find *Craibia brevicaudata*, Rhodesian Ironwood.

- **umzithi**: False Tamboti, *Cleistanthus schlechteri*. Is this *umuzi* (homestead) + *-thi*? Boon says "wood used for hut-building" (2010: 232);
- **inkubathi**: Lavender Croton, *Croton gratissimus* ? < *inkuba* (dungbeetle)?;
- **umfongothi**: River Macaranga, *Macaranga capensis* is also **umfongafongo** (Boon 2010) and **umfongafonga, umfongofongo** (Doke and Vilakazi 1958) and Sausage Tree, *Kigelia africana* (also **umvongothi** and **umbongothi**));
- **umhluthi**: Red-Beech, *Protorhus longifolia* and also **uhlangothi, umhlakothi**;
- **isikhungathi**: White Mangrove, *Avicennia marina*;
- **uzwathi**: Forest Bell-Bush, *Mackaya bella*;
- **umthongothi**: Spiny-Gardenia, *Hyperacanthus amoenus*; and
- **umkhambathi**: Paperbark Acacia, *Acacia sieberiana*, also **umkhamba** (see Plate 12).

There are two rather curious examples:
- **umhlalamithi:** Common Hard-Leaf, *Phylica paniculata*, with the plural form *imithi* rather than the singular form *umuthi*; and
- **umuthinzima**: False Soap-Berry (hard tree), *Pancovia golungensis*, where *umuthi* occurs as the first part of the compound rather than the second.

The following examples are from Pooley (1998), so these are wild flowers rather than trees and confirm Van Warmelo's suggestion that in these compounds *umuthi* has the sense of 'plant' rather than 'tree':
- **imbathi**: Stinging Nettle, *Urtica urens* annual herb and widespread weed with the Swazi name *isibathi*;
- **indlolothi**: Large Yellow Moraea, **Moraea spathulata* and Yellow Tulip, *Homeria pallida*;
- **inhluthi yotshani**: Death Orchid, *Habenaria dives*;
- *isicakathi*: Xhosa name for Cape Smilax,[20] *Asparagus asparagoides*;
- **ugonothi**: Climbing Bamboo, *Flagellaria guineensis*;
- **ukhathimuthi**: Soccerball Pachycarpus, *Pachycarpus appendiculatus* and Tongued Pachycarpus, *Pachycarpus dealbatus*;

---

20. This vernacular name given by Pooley (1998: 514) for *Asparagus asparagoides* should not be confused with the genus name *Smilax* in the family *Smilacaceae*.

- **umdlandlathi**: Traveller's Joy, *Clematis brachiata*; and
- **usithathi**: Wild Clover, *Trifolium burchellianum*.

I have been convinced for many years that the Zulu tree name **umthombothi** (Tamboti, *Spirostachys africana*) was a compound of *umthombo* (fountain, spring) + [*umu*]*thi* (tree), the name indicating that the tree grew near underground water. But Van Warmelo offers an intriguing alternative:

> *Mutomboti* (Venda) is the tree **Canthium mundianum*, a very hard wood out of which clubs are carved, actually *mu-tombo-ti*, i.e. *-tombo* 'stone' + *-ti* 'tree', therefore 'stone-tree', wood as hard as stone. Though the root *-tombo* 'stone' is now not known in Tsonga and Zulu/Xhosa, we still find in Swazi *um-tfombo-tsi*, Zulu *um-thombo-thi* '*Spirostachys africanus*', which is also a very hard wood (1976: 93).

Wild (1952: 39) gives *muTomboti* as the Ndau name of *Spirostachys africana*[21] and *muTombo* (i.e., without the *-ti*) as the Shona name for Camel's Foot (*Piliostigma thoningii*[22] or Monkeybread in Southern Rhodesia as it was).

Other tree names ending with *-ti* from the different Shona dialects spoken in Zimbabwe (all from Wild 1952) include:

- *riTsanzwiti*: Manyika, *Freylinia tropica*;
- *muTsatsati*: Shona = *Lannea edulis*, Karanga, Zezuru = *Faurea speciosa & F. saligna*;
- *chiSosoti*: Shona, *Grewia inaequilatera*;
- *muShati*: Karanga, *Erythrophleum africanum*;
- *muRuwati*: Shona, **Diospyros nummularia*;
- *muRwiti*: Shona, *Combretum imberbe*;
- *muPfumoti*: Ndau, **Piptadenia buchananii*;
- *chiPambati*: Manyika, **Euclea multiflora* and *Buddleja salviifolia*;
- *Gwiniti* and *Guneti*: both Ndau, *Celtis durandii*; and
- *Gombati*: Ndau, *Erythrina abyssinica*.

---

21. In the then Southern Rhodesia of 1952, this tree went by the alternative English common names African Sandalwood and Tambootie Wood (Wild 1952: 126).
22. Coates Palgrave merely says "the wood makes a good fuel but other than this it has little value". He spells the specific name *thoningii*, which would appear to be the correct form (1977: 285).

Van Warmelo (1976: 94) gives a number of Sotho plant names ending with -*re* and a number of Venda plant names ending with -*ri*, of which the following are a few examples.[23]

From Sotho:

- *mohlwelere*: **Combretum suluense* and other spp. of *Combretum*;
- *motlhabare*: **Lachnopylis floribunda*;
- *mogotlhore*: **Pygeum africanum*; and
- *mogokare*: *Salix woodii*.

From Venda:

- *tshidiri*: **Grumilea capensis*;
- *mudzwiri*: sp. of hardwood;
- *mutasiri*: *Rhus transvaalensis*; and
- *musiri*: **Olea foveolata*.

As we saw earlier, Afrikaans also frequently adds -*boom* (tree) or -*hout* (wood) to single words for various species of tree to form another single word that is a compound: *denneboom* instead of *den* (pine, fir), *lariksboom* for *lariks* (larch), *sederboom* or *sederhout* for *seder* (cedar). *Wilg* (willow) alternates with *wilgeboom* and *kersie* (cherry) with *kersieboom*. Some tree names only occur in this form: *beukeboom* or *beukehout* (beech), *essenhout* (ash) and *berkeboom* (birch).

Using an element meaning 'tree', 'bush' or 'plant' in a plant name is by no means restricted to vernacular naming systems. The Linnaean botanical nomenclature system also does it. The examples that follow are all from Boon (2010):

- *Calodendrum*: < Gk *kalos* (beautiful) + *dendron* (tree) referring to attractive flowers);
- *Clerodendrum*: < Gk *kleros* (chance) + *dendron*. Plants in the genus have widely varying medicinal properties;
- *Elaeodendron*: < Gk *elaia* (olive) + *dendron*. The fruit are oily and look like olives;
- *Toxicodendron:* < Gk *toxikos* & Lat. *toxicum* (poison) + *dendron*. The sap contains a powerful allergen;

---

23.   -*re* and -*ri* are cognates with the -*thi* of the Nguni languages.

- *Acacia melanoxylon*: < Gk *melanos* (black) + *xylon* (wood). Australian Blackwood. See also *Dalbergia melanoxylon*;
- *Erythroxylum*: < Gk *erythros* (red) + *xylon*. In some spp. the wood is red;
- *Mystroxylon*: < Gk *mystron* (spoon) + *xylon* (not explained);
- *Zanthoxylum*: < Gk *xanthos* (yellow) + *xylon*. Yellow heartwood and roots;
- *Ptaeroxylon*: < Gk *ptairein* (sneeze) + *xylon*. Sneezewood, *nieshout*;
- *Phytolacca*: < Gk *phyton* (plant) + Hindi *lakh* (dye). The fruits stain badly; and
- *Pygmaeothamnus*: < Gk *pugmaios* (dwarf) + *thamnos* (bush).

We should note, though, that the botanical names *Commiphora woodii, Calpurnia woodii, Combretum woodii* and several others with the specific name *woodii*, do not belong to the category above. The 'wood' in these names commemorates John Medley Wood (1827–1915), farmer, trader and botanist of KwaZulu-Natal, curator of the Durban Botanical Gardens from 1882 and founder of the Natal Herbarium.

There are a number of interesting examples of botanical names in which an embedded anthroponym is added to a term for 'tree' or 'plant'. Again, the examples are from Boon (2010):

- *Robsonodendron eucleiforme*:[24] False Silky-Bark "after Norman Robson of the British Museum + Gk *dendron* 'tree' so 'Robson's tree'" (Boon 2010: 314);
- *Englerodaphne ovalifolia*: Broad-Leaf Fibre-Bush after H.G. Adolf Engler, botanist of Berlin + *daphne* (Gk = laurel) i.e., Engler's laurel. See also *Englerophytum magalismontanum* (stemfruit), where *Englerophytum* is from Engler and *phyton* (plant). An earlier synonym is *Bequaertiodendron*[25] *magalismontanum*; and
- *Dahlgrenodendron*: < Swedish botanist R.M.T. Dahlgren (1932–1987) + *dendron*.

---

24. Beaumont (personal communication) regards this as an unresolved name. Some, she says, suggest that it is *Cassine eucleiformis*.
25. For Joseph Charles Corneille Bequaert (1886–1982), Belgian-born American botanist who worked in the Belgian Congo.

Morphological variations

As part of an oral knowledge system, the Zulu name of a plant may vary in form. I am not talking about different names for the same plant here, but different forms of the same name. For example, the Zulu name of the plant *Anemone caffra* is variously given as **umantingwe**, **intingwe** or **untingwe**. Each of these can be considered 'correct'; it is not the case that one is correct and the others are misspellings.[26]

We have already looked in detail at the considerable number of potential forms for the name **umfana-kanosihlanja** (the Arum Lily, *Stylochiton natalensis*). Let us now look more generally at morphological variation in Zulu plant names, starting with the various name-forming morphemes found in Zulu.

Name formatives in Zuḷu

We saw above how the formatives *-ma-* and *-no-* are used to create names in Zulu. Less commonly found formatives are the two prefixes *so-* and *sa-*, and the suffix *-se*.

So- has the meaning 'male' if found in personal names such as *uSomfana* (with *umfana* or boy). It is also often translated as 'father of' in names such as *uSobantu* (father of the people, the Zulu name for Bishop John Colenso), and *uSomandla* (father of strength) and *uSokulunga* (father of correctness), both names for God, with the alternative translations 'The Almighty' and 'The All-righteousness'. The meaning 'male' is lost in such words as *usompempe* (referee < *impempe*, whistle) and *usomabhizinisi* (businessman, businesswoman < *ibhizinisi*, business). *So-* is also found in such species names as *usokhuni* (species of sea animal < *ukhuni*, firewood) and *usomheshe* (species of grey hawk < *uheshe*, hawk).[27]

Sa- has the meaning 'something like' or 'something to do with', as in *isamuntu* (ghost, spectre, something like a human < *umuntu*, human) and *isangoma* (diviner, i.e., someone who has something to do with *ingoma*, song). We shall give several examples of plant names incorporating *sa-* later.

-Se is an unproductive morpheme (i.e., one found only in historical forms and not used for creating new names or nouns today). It is found in many Zulu

---

26. Incidentally, if you can't make up your mind which one of the three variants to use, the herbalist's name for this plant, **umanzamnyama** (black water), can be used instead.

27. **usomheshe** also refers to a "species of small-leaved sweet-potato, with purplish tubers" (Doke and Vilakazi 1958: 764).

clan names such as uMdlalose (< *umdlalo*, game) and uShangase (< *shánga*, of wandering about). It is found in some bird names, such as *unkombose* (Namaqua dove < *inkombo*, a pointing < *khomba*, to point) and *ukholwase* (flamingo < *kholwa*, be believed).

An interesting example of both *so-* and *-se* coming together in one plant name is **usombombose**, which according to Doke and Vilakazi is a "species of small-leaved sweet-potato" (1958: 764) and derived from **umbombo** (sp. of creeping plant, used for sprinkling medicine). It is also interesting to note that the only other example I have of a Zulu plant name using the prefix *so-* also refers to a small-leaved sweet potato (see footnote 27).

In the examples that follow there are phonological as well as morphological changes. For example, **umbhangabhanga, umbhongobhongo, umbhengabhenga** and **umbhengebhenge** are all forms that Doke and Vilakazi give for *Trema bracteolata*.[28] All the following examples, unless there is specific indication to the contrary, are from Doke and Vilakazi (1958) and the numbers in brackets refer to pages in their *Dictionary*.

In the following sets of names, variation is caused by the presence or absence of *-ma-* (as well as other minor changes). Doke and Vilakazi (1958) give both **umcumane** and **umancumane** as names for the "ornamental forest tree, *Alberta magna*" (128). They cross-reference two variants as "**umadoye** (o-) species of plant with sweet-smelling flowers, *Asclepias albens*, cf. *i(li) doyi*" (475) and "**idoyi** name applied to *Asclepias albens*, and similar species eaten by natives cf. *umadoye*" (168). A third variant, **udoye**, however, refers to a different species, a "small bush, *Maesa alnifolia*" (168). Both presence or absence of *-ma-*, and what looks like a curious passive form of a noun,[29] is found in the names Doke and Vilakazi give (402, 478) for a species of *Asclepias*: **isikhondwe** (sp. of plant with edible root), **isikhonde** (*Asclepias multicaulis*, roots edible) and **umakhondwe** (sp. of *Asclepias* plant). The same apparent passive form of a noun is found in the variant **umsimbithwa** for the tree **umsimbithi** discussed in considerable detail above.

Note that the *-ma-* in the following pair is not the same *-ma-* previously discussed: **umafusi** and **umafusini**, both of which Doke and Vilakazi give for "1. sp. of iris lily, *Aristea ecklonii*; 2. Natal red-top grass, **Rhynchelytrum* *setifolium*" (475). The base of these two forms is the noun *ifusi* (fallow land), in the plural form *amafusi*, but placed in class 3a as **umafusi**. In the form **umafusini**, the base is the locative *emafusini* (in the fallow lands), likewise moved to class 3a. We see the same movement of a noun from its home noun class to class 3a while still retaining a short form of the original noun class prefix in the class 3a noun *usithundu* (a type of love charm emetic), derived from the class 7 noun *isithundu* (medicine calculated to bring prosperity).

In the following set, the presence or absence of *-no-* causes the variation:

- **unomalenjane** and **amalenjane**: the common Purslane, *Portulaca oleracea*); and
- **unomazele** and **izele** (plural *amazele*): **Grumilea capensis*, an undershrub of the mist belt forests.

In the following set of names, *-sa-* plays a role in creating variations. The name **isankuntshane**, referring to "*Ophioglossum reticulatum*, the adder-tongue fern, esteemed as a vegetable" (Doke and Vilakazi 1958: 12), has the variant form **isinkuntshane**. We find the same variation in **isagude** and **isigude** with both forms referring to the Wild Banana **Strelitzia augusta*. It is interesting that Doke and Vilakazi refer to **isakhwali** as a "<u>dialectal</u> variant of *isikhwali*" (1958: 5).[30] They give **isakhwali** as a "species of tuberous veld plant" and **isikhwali** as "1. species of tuberous veld plant, **Vigna triloba, V. vexillata, *V. glabra* (dialectal variant *isakhwali*); 2. certain other papilionaceous climbers used in feverish conditions and as love charms" (1958: 420). When **isikhwali** is followed by **sasolwandle** (of the sea), it refers specifically to the Gloriosa Lily. The plant name **isikhwali** in its plural form has found its way into the Zulu saying *ukuphanda izikhwali* (to draw out a job a long time, or literally dig out tubers).

The noun stem *-ncasha* appears in two guises: as **isancasha**, which Doke and Vilakazi give as a name for a "species of tree, **Schotia transvaalensis*, the Boer-bean" (1958: 9); and as **umncashane**, referring to "*Cryptocarya*, a wild laurel tree" (528). The presence and absence of the diminutive *-ane* here is echoed in the pair **isibhaha** and **isibhahane**. **Isibhaha** is given as the name for the Fever Tree, **Warburgia breyeri* "whose very hot and ginger-like root-bark is used for malarial fever and as an expectorant (< v. *bhaha* 'rage in anger' > *isibhahane*)" and **isibhahane** refers to "*Wahlenbergia undulata*, a bitter emetic" (20).

---

30. Author's emphasis in quotation.

Doke and Vilakazi (1958: 57, 58) give two forms of the name for the Stinging Nettle. One is the simple-stemmed name **imbaba**, the other is the same stem with both *-azi* and *-ane* added to form **imbabazane**.

Sa-, *-ma-* and change of noun class are all responsible for a complex set of names relating to wild carrots and other plants:

- **isaqathe**: a species of edible wild carrot, growing as a runner above ground, cf. *isanqante* (Doke and Vilakazi 1958: 13);
- **isanqante**: "1. *Asclepiadacea*, species of edible wild carrot [cf. **isaqatha**, **inqantu**][31]; 2. certain kinds of *Schizoglossum* cf. **umanqante**" (12);[32]
- **umanqante**: < *inqante*: "species of plant with edible root, *Schizoglossum robustum, S. punctatum, *S.woodii*" (484);
- **inqante**: "species of plant of the Carrion-flower family more commonly heard as *umanqante*" (591);
- **iqathe**: fruit of certain species of spreading veld shrubs, in the plural **amaqathe** used of the whole shrub *Pachystigma pygmaeum*; and
- **isiqathe**: "place overgrown with *amaqathe* plants (cf. *isaqathe*)" (13).

The following show mainly phonological differences with some apparent affixes:

- **umbhanga**: "species of tree found in the Ngoye forest" (34);
- **umbhangabhanga**: "term applied to certain quickly growing trees or shrubs (cf. **umbhongobhonga**): 1. *Trema bracteolata*, a river tree; 2. *Canthium queinzii*, quickly growing river bush; 3. *Cricus*[33] *lanceolatus*, purple-headed thistle; 4. *Berkheya grossa*,[34] species of thistle" (34);
- **ubhangabhu**: "species of tree. *Macaranga capensis*" (24);
- **umbhangazi**: "name used for various quickly growing trees, esp. *Albizia mossambicensis*" (24);
- **umbhengabhenga**: *Trema bracteolata* tree;
- **umbhengebhenge**: *Trema bracteolata* tree; and
- **umbhengele**: "Certain species of tree: 1. *Trema bracteolata* tree; 2. *Cussonia umbellifera*; 3. *Macaranga capensis*" (33).

---

31. Doke and Vilakazi say "cf. inqantu", but *inqantu* does not have a separate entry.
32. Beaumont points out (personal communication) that these plants have now been placed in the family *Apocynaceae*.
33. Doke and Vilakazi give this as 'Cricus'. It took considerable effort to track this down as the correct 'Cnicus'.
34. We have been unable to trace this plant.

In the following six sets of name variations, the difference is entirely phonological. There is no structural difference between them:

- **ibhada**: "species of veld plant, cf. **ibhade** species of veld plant, the white underskin of whose leaves were stripped for fringes and body ornaments. Term applied to several plants: *Buphane*[35] *disticha*, **Helichrysum leiopodium*, *H. appendiculatum*, *H. cephaloideum*" (18);
- **umagunda/umaginda**: Assegai-Wood Tree, *Curtisia faginea (476);
- **umganu**, **umgamu** and **umgani**: all refer to **Sclerocarya caffra*, the Maroela (231, 232);
- **umembeso**, **umembezo** and **umembeza**: all refer to a species of small shrub (185);
- **umbhemethu**: **Gardenia neuberia* plant (v.l. *isibhembhedu*) (33) and **isibhembhedu**, "name applied to five different species of shrub or trees with rectangular cross-branching, viz. 1. *Anastrabe integerrima*, a river timber tree; 2. **Gardenia neuberia* (v.l. *umbhembethu*); 3. *Plectroniella armata*; 4. **Canthium obovatum*; *C. ciliatum*" (32); and
- **umbanda**: "species of tree, cf. *umbandu*" cf. **umbandu**: species of tree, *Lonchocarpus capassa*" (63).

There is something about the following two groups suggesting that in some much earlier form of Nguni *bub- and *bung- could have been proto forms:[36]

- **u(lu)bubu**: sp. of small scandent shrub, *Choristylis rhamnoides* (cf. **u(lu)mbhimbi**, **unobubu**);
- **u(lu)bubane**: creeping grass found on river banks;
- **i(li)bubathe**: species of herb *Urtica urens*, v.l. *ibubazi*; and
- **ibubazi**: *Urtica urens* (Doke and Vilakazi 1958: 87).
- **umbungaze**: *Berkheya grossa* plant, sp. of Thistle;
- **umbungashe**: *Lichtensteinia interrupta* herb, Wild Carrot;
- **ibunge**: sp. of plant;
- **umbunge**: sp. of tree, *Greyia sutherlandii*; and
- **ibungu**: "1. sp. of river grass or rush, *Cyperus papyrus*; 2. veld grass when still young; 3. **Landolphia capensis*" (Doke and Vilakazi 1958: 92).

---

35. This is how Doke and Vilakazi give the name. The correct form is *Boophone*.
36. These are hypothetical or postulated forms, hence the initial asterisk.

Our last example is a curious one. It does not refer to morphological or phonological variation as in the examples above, but rather a simple-stemmed noun to which has been added another noun to form a compound noun. The Zulu word **imbuya** for the *Amaranthus* species Pigweed is well known. A regularly eaten species of plant, it is tinned and marketed in supermarkets under the name Marogo. Doke and Vilakazi (1958: 95) say **imbuya** is a Zulu development from the Ur-Bantu word *vuya* (herb). So it is simultaneously (1) a semantically wide-ranging word equivalent in English to words like 'tree', 'bush' and 'grass'; (2) a very old word; and (3) a word widely spread among Bantu languages, otherwise Meinhof would not have been able to postulate an Ur-Bantu form.[37] Doke and Vilakazi also give the plant name **imbuyabathwa** (*imbuya* + *abathwa*, Bushmen) as "Prickly pig-weed, *Amaranthus spinosus*, inedible". In other words, the name for the Khoisan or Bushmen people is used as a negation of edibility. That is what I find curious.

Words in all languages have both shape and meaning. In this chapter we have looked at the shape (structure) of Zulu plant names. Now we go on to look at meaning.

---

37. Carl Meinhof (1857–1944) was a German philologist who studied the Bantu languages at the beginning of the twentieth century and postulated root forms for a large number of cognate words.

4 The semantics of plant names

"My *name* is Alice, but – "

"It's a stupid name enough!" Humpty Dumpty interrupted impatiently. "What does it mean?"

"*Must* a name mean something?" Alice asked doubtfully.

"Of course it must," Humpty Dumpty said with a short laugh: "my name means the shape I am – and a good handsome shape it is, too. With a name like yours, you might be any shape, almost."[1]

Introduction

Commenting on the extract above, Gardner says:

In real life proper names seldom have a meaning other than the fact that they denote an individual object, whereas other words have general universal meanings (1960: 263).[2]

The debate about whether or not proper names have meaning has been going on a long time. Some linguists say proper names have no meaning at all, others

---

1. This extract is from Lewis Carroll's *Through the Looking-Glass and What Alice Found There*, first published by Macmillan in 1871.
2. Martin Gardner's *The Annotated Alice* (1960) contains his edited version of Carroll's *Alice's Adventures in Wonderland* and *Through the Looking-Glass and What Alice Found There*.

say they are the most meaningful of all words. It is not the role of this book on Zulu botanical nomenclature to go into this debate; indeed as this book is not about proper names at all, it would be pointless. But the quotations above from Alice, Humpty Dumpty and Martin Gardner raise questions about meaning that <u>are</u> very relevant to this book, and we need to look at the issues raised.

All words have meaning, but here I am going to talk only about nouns, and leave out all the other parts of speech. 'Traditional' school grammars for English talked of 'proper nouns' and 'common nouns'. Proper nouns are 'names', like 'Alice', 'Humpty Dumpty', 'Texas', 'Durban', 'Mount Everest' and 'The Mississippi'. Common nouns, sometimes called 'appellatives', were in the same traditional grammars sub-divided into 'concrete' (e.g., dog, washing machine, thunderstorm) and 'abstract' (e.g., peace, democracy). The Zulu plant 'names' in this book are all 'common nouns', and all 'concrete'.

Gardner's comment is somewhat misleading. He says that proper names denote whereas other words have general universal meanings. In fact the primary function of <u>all</u> nouns, whether proper names or not, is to denote. All nouns are words that refer to 'real life' entities, states and conditions. 'Alice' is used to denote, or refer to, any female with the name Alice.[3] 'Humpty Dumpty' is the name of an egg-shaped character in a book. 'Texas' refers to one of the states of the United States. The name 'Mount Everest' singles out the highest mountain in the world. 'Durban' denotes a large port city on the KwaZulu-Natal coast. 'Dog' is used to refer to domestic canine quadrupeds. The word 'peace' denotes the absence of fighting, violence or intrusive sound. And so on.

We can call these 'referential meanings' or 'denotative meanings' and although these are almost always the primary meanings of words, they are by no means the only meanings that a word can have.

Some words have many different references. 'Spring' for example, could mean 'leap in the air', 'the season of the year between winter and summer', 'water bubbling up from the earth' or a 'device made of tightly coiled wire'. The American humorist James Thurber, in an amusing story about word games,[4] gives fourteen different meaning for the word 'dog'.

As we noted in Chapter 2, Zulu plant names often refer to several different species, so they can also be said to have a number of different references.

---

3. Or even, in exceptional circumstances, to 'non-females' named Alice as in American rock star Alice Cooper.
4. Thurber 1962: 129.

According to Pooley (1998: 296), the Zulu plant name **inqwambi** refers only to the species *Diospyros galpinii* (Dwarf Hairy Jackalberry). We can say, then, that the referential meaning of this word is 'the plant species *Diospyros galpinii*' or alternatively 'any specimen of the species *Diospyros galpinii*'. When the word is used in the plural form – **izinqwambi** – it no longer refers to the species as a whole and has the sole meaning 'two or more specimens of *Diospyros galpinii*'. Pooley (1998: 222, 448) gives the name **idangabane** to refer to both *Commelina africana* (Yellow Commelina) and *Commelina benghalensis* (Benghal Commelina), which to the thinking of Western botanical taxonomy means the word has two references. In Zulu folk-taxonomy these two Commelinas would be considered to be the single plant **idangabane** with only one referential meaning. The same is surely true of **icena** (Pooley 1998: 34, 342) for *Aloe parvibracteata* and *Aloe greenii*. And although Pooley uses the Zulu **icacane** to refer to five different species (1998: 26, 28, 30, 226, 508), they are all species of *Kniphofia*. Again, in Zulu folk-taxonomy there would only be one reference.

On the other hand, **idololenkonyane** clearly has many references in both the Western and Zulu taxonomies with Pooley (1998: 282, 374, 376, 490, 526, 576) using this word to denote all the following species:

- *Cyphostemma natalitium*
- *Rumex lanceolatus* (Smooth Dock)
- *Persicaria lapathifolia* (Spotted Knotweed)
- *Barleria obtusa* (Bush Violet)
- *Rumex crispus* (Curly Rumex/Narrow-Leaved Dock)
- *Hydnora africana* (Warty Jackal Food)

Words may also have 'underlying' meanings. **Idololenkonyane** above has six different referential (denotative) meanings, but also the underlying meaning 'knee of the calf' (< *idolo* (knee) + *la* (of) + *inkonyane* (calf)). There are several Zulu plant names similar to this, such as:

- **ulimilwenyathi** (buffalo's tongue): *Berkheya setifera*, for which Pooley (1998: 336) gives the English name Buffalo-Tongue Berkheya and makes a neat three-way link between the Zulu, Latin and English names;
- **umsilawengwe** (leopard's tail): "species of shrub, **Gnidia kraussii*" (Doke and Vilakazi 1958: 755) (see Frontispiece);
- **unyawolwendlovu** (elephant's foot): *Dioscorea cotinifolia*. Pooley (1998: 108) gives the English name here as Wild Yam, but the Afrikaans

name *olifantsvoet* make the same kind of three-way nomenclatural link as we saw with **ulimilwenyathi**; and

- **amabelejongosi** (young maiden's breasts): *Polystachya ottoniana* and a number of other species (Pooley 1998: 124, 242, 246, 372).

Many English vernacular names of wild flowers have the same type of underlying meaning as the Zulu names above:

- Lady's Mantle **Alchemilla vulgaris* (Stevens 2008: 33);
- Bullock's Heart, another name for the Custard Apple (*Collins Dictionary of the English Language* 1986: 362);
- Dog's Mercury, *Mercurialis perennis* (Stevens 2008: 20); and
- Hound's Tongue, *Cynoglossum officinale*.[5]

In the last example, the common name and the generic name both derive from the same underlying concept, with the generic name taken from Greek *kyon* (dog) and *glossa* (tongue) (Stevens 2008: 31). A similar situation exists for Goosefoot (*Chenopodium* spp.). Pooley (1998: 526) says that *Chenopodium* is derived from the Greek *cheno* (goose) + *podion* (little foot) in reference to the shape of the leaves of this genus.[6]

I use the term 'lexical meanings' for these underlying meanings, because they are the meanings of the elements of the names as lexical items: i.e., as words found in a dictionary. Doke and Vilakazi's *Zulu-English Dictionary* usually gives both the referential/denotative as well as the underlying lexical meanings of plant names, as in the following four entries:

unyawolwenkuku . . . n. [< u(lu)nyawo + poss. inkuku, lit. fowl's foot]
1. species of long-shaped edible tuber . . .
2. *Pelargonium aconitifolium* [= **Pelargonium aconitophyllum*], with leaves sometime shaped like a fowl's foot.[7]

---

5. Beaumont (personal communication) suggests that 'Lady's mantle' refers to the leaf shape, similar to Elizabethan lace collars, that 'Hound's tongue' is also based on leaf shape, and that 'Dog's mercury' is a typical example of the use of the word 'dog' in plant names to indicate something very common.
6. *Collins Dictionary of the English Language* tells us that the genus *Chenopodium* is also known by the English vernacular names Good King Henry and Fat Hen (1986: 656).
7. Note how Doke and Vilakazi link the referential meaning to the underlying meaning with this explanation. See page vi.

utshwalabezinyoni . . . n. [< utshwala + poss. izinyoni, lit. birds' beer]
1. Red dagga, *Leonotis leonurus* . . .
2. Other species of ornithophilous[8] flowers . . .

umnukambiba . . . n. [< nuka + imbiba, lit. 'what smells out the field rat']
Species of small bush, *Myaris* or *Clausena inaequalis*, with strong-smelling, inflammable leaves . . .[9]

umdlampunzi . . . n. [< dla + impunzi, lit. 'what the duiker eats']
Species of fruit tree, **Sapium reticulatum*.

Relationship between underlying meanings and plant itself or Zulu cultural usage of plant

Humpty Dumpty said his name referred to his ovoid shape. As we shall see in Chapter 6, Zulu plant names may also refer to shape, as well as habitat, growth type, appearance, taste and smell. The names **amabelejongosi** and **amabelentombi** (both meaning maiden's breasts), mentioned earlier, can refer to the shape of individual flowers as well as to knobby growths on the trunk of a tree (see illustration on page 79).

These underlying meanings, and their relationship to the plant itself, or to the Zulu cultural use of the plant, are explored in great detail later in this book, so I shall give only a few more examples here:

- **Euclea daphnoides* (Boon 2010: 466) is known as **idungamuzi** (what disturbs the home) and **ichithamuzi** (what disperses the home) and extracts from this tree are used with evil intent by sorcerers who wish to disturb the harmony of a home. For this reason, as Boon points out, the wood is avoided for fuel in Zululand;
- *Aloe barberae* (Boon 2010: 54) has the name **umpondonde** (long horns), a reference to the shape of its leaves;
- *Cryptocarya woodii* (Boon 2010: 108) has as one of its Zulu names **isilindangulube** (what the pig waits for). Doke and Vilakazi tell us that this tree "bears a fruit liked by wild pigs" (1958: 458);

---

8. *Collins Dictionary of the English Language* does not recognise this word. Beaumont (personal communication) explains it as the term used to describe pollination by birds.
9. *Myaris inaequalis* is an earlier synonym of *Clausena inaequalis* (Beaumont, personal communication).

The cone-shaped knobs on the trunks of the Knobwood trees *Zanthoxylum capense* and *Z. davyi* give rise to the Zulu name **amabelentombi** (maiden's breasts) (see pages 78 and 142) (drawing by Angela Beaumont).

- *Calpurnia aurea* goes by the Zulu name **umkhiphampethu** (what takes out the maggot). As Boon explains, "infusion used against maggot-infested sores in cattle . . . crushed roots used against lice" (2010: 148);
- *Cyrtanthus breviflorus* (Pooley 1998: 232) is used in Zulu society as a love charm. The name **uvelabahleke** (appear and they smile) refers to use of this plant to make a girl attractive to men; and
- *Ansellia africana* is eaten by monkeys (Pooley 1998: 242). This is reflected in the Zulu name **imfeyenkawu** (sweet-reed of the monkey).

Sometimes more than one vernacular name coincides in underlying meaning. This is the case with **Kalanchoe thyrsiflora* (Pooley 1998: 254) where the Zulu name is **utshwalabenyoni** (bird's beer). One of two English names is White Bird's Brandy and one of three Afrikaans names is *voëlbrandewyn*.

The Drakensberg flower *Ornithogalum juncifolium* has a generic based on Greek *ornis* (bird) and *gala* (milk) (Pooley 1988: 98). This is clearly the same idea as **utshwalabenyoni** above and **utshwalabezinyoni** (birds' beer), the Zulu name for the flower *Leonotis leonurus*. Note that in **utshwalabenyoni** only one bird 'drinks the beer' whereas in **utshwalabezinyoni** several birds are involved.

Connotative and associative meanings

Let us return to Alice's conversation with Humpty Dumpty, quoted at the start of this chapter. Humpty Dumpty says "my name means the shape I am". To those familiar with Tenniel's illustrations in Carroll's two Alice stories, this means 'egg-shaped' (Gardner 1960: 264). Such a meaning is sometimes referred to as a 'connotative meaning' and sometimes as an 'associative meaning'. Familiarity with the Tenniel picture will allow 'Humpty Dumpty' to connote 'egg-shaped'; or, put another way, Carroll fans will associate eggs with the name Humpty Dumpty. Of course, those not familiar with Carroll and his illustrator Tenniel, will have neither connotative nor associative meanings for the name 'Humpty Dumpty'.

I personally prefer to distinguish between 'connotative' and 'associative' meanings. To me, connotations are in the word itself, associations about what is known of the referent. Let us take two names for the same city, 'eThekwini' and 'Durban'. Both have the same referential or denotative meaning: 'large port city on east coast of KwaZulu-Natal in South Africa'.[10] 'Ethekwini' is based on the locative form of the noun *itheku* (bay, lagoon)[11] and so has the underlying meaning 'at the bay' or 'at the lagoon'. The connotations here are that this must be a place at a river mouth on a sea shore. You could not give this name to a city elsewhere than in such a location. Nor could you give this name to a mountain or a forest. You could not even give it to other water features such as a stream, a spring or a waterfall. The connotations would not permit such referents.

On the other hand, 'Durban' has no such connotations. In terms of its etymology, it is derived from the name of a one-time Governor of the Cape, Sir Benjamin D'Urban, but this knowledge gives us no such locational information as 'eThekwini' does. The associative meanings of Durban will vary from person to person, based on their experiences and what they know of Durban. For some, the associations will be of fun-filled holidays on sunny, sandy beaches; for others sports and games played in the city; for yet others, the name will mean ships and a harbour.

---

10. For 'Durban', *Collins Dictionary of the English Language* (1986: 475) gives: "a port in E South Africa, in E Natal on the Indian Ocean; University of Natal (1909); resort and industrial centre, with oil refineries, shipbuilding yards, etc. Pop.: 505 963 (1980)". This entry is equally valid for 'eThekwini'.
11. For those interested in alternative meanings, see Koopman (2009).

Zulu personal names frequently have connotative meanings. The girl's name uNtombifuthi (< *intombi* (girl) + *futhi* (again)) connotes a situation where a daughter is born and then instead of the hoped-for son, another daughter (girl again) is born. This name can only be given to the second-oldest girl. Similarly, the boy's name uMfanufikile (< *umfana* (boy) + *ufikile* (he has arrived)) refers to a situation where a boy is finally born after a number of girls. Note that these connotative meanings rely on both the underlying lexical meanings as well as knowledge of the aetiology (underlying reasons) for Zulu name-giving.

Sometimes knowledge of specific family circumstances is necessary to get the full connotations of a name. For example, let us take the boy's name uNkomokazikho. This is derived from [*izi*]*nkomo* (cattle) and *kazikho* (they are not present). The underlying meaning is thus 'there are no cattle'. I have recorded this name six times in research into Zulu personal names and in each case was able to ask a parent why the name had been given. In three cases, the boy was born shortly after or during a period of extended drought that had brought death to many of the father's cattle. In the other three cases, the boy so named was one at the end of many boys in a boys-only family. Zulu fathers contribute cattle to their sons' marriage dowries and get others back when their daughters marry. This is why Zulu parents like to balance boys with girls.[12] When there are only boys, cattle go out and none come in. For this reason, a common name for a girl born after many boys is uZibuyile (they [cattle] have come back).

Note here the same lexical meaning has two quite distinct connotative meanings and only knowledge of the circumstances in each case will give this full meaning. I can imagine that in other cases with the same name (uNkomokazikho), stories can be found telling of cattle that have died of disease, or were impounded by officials, sold off to pay a debt, struck by lightning and so on.

A similar situation where name interpretation is related to specific circumstances arose when my Dalmatian dog was given the Zulu name uBahlangene by a Zulu-speaking employee of the staff club at the then University of Natal in Pietermaritzburg. The name literally means 'they are gathered together' (the lexical, or underlying meaning). This name is a common Zulu dog's name, given to a dog when the owner wishes to use the name out aloud to let his neighbours know that he is aware of the fact that they have

---

12. See Koopman (2002: 35, 36).

gathered together against him.[13] Suspicions of witchcraft and tensions between neighbours are commonly associated with this name. As I did not understand why the person concerned had given my dog this name,[14] I asked him for his reason. His explanation was that as the University Club was one of the few places in town where black and white people could gather together and drink in harmony (this was in the apartheid years when black and white people were segregated), he had named my black-and-white spotted dog 'They-are-gathered-together'.

Another example of a name where the deeper, underlying reason is not apparent in the literal (lexical) meaning is the Zulu plant name **intombikayibhinci**, which Boon (2010: 78) gives for the fig *Ficus sur*. The literal meaning of this name is 'the girl does not wear clothes', from *intombi* (girl) and the negative form of *bhinca* (wear traditional clothing). Without specialised knowledge it would be difficult, if not impossible, to guess how this literal meaning could relate to a plant. My colleague Ndela Ntshangase,[15] who is familiar with the way certain parts of this tree are used for love charms, says that "this love charm is so powerful that when the girl takes it and then sees you, she does not even wait to get dressed, or if she is dressed, she flings off all her clothes when she sees you".

Many Zulu plant names have such connotative and associative meanings. For instance, a plant name containing the element *hlaba* (prick, stab) will speak of thorns. All of the following are thorny or prickly: **umhlaba**, several species of large aloes; **umhlababa**, a species of thorny shrub; **umhlabahlungulu** (what stabs the crow), a species of *Phyllanthus* a tree found in Ngome forest; and **isihlabamakhonjane** (what stabs the baboons), a species of thistle. The verbs *bamba* (catch hold of) and *bopha* (arrest, tie up) have similar connotations, as in **umbambampala** (what catches the impala), a species of bush with hook thorns; **umbambangwe** (what catches the leopard), a species of large tree with great thorns; and **umbophanyamazane** (what arrests the buck), a species of tree that forms impenetrable thickets.

The Zulu name **imbune** (Sensitive Plant) refers to the tree *Mimosa pigra* (Boon 2010: 198). The name is derived from the verb *buna* (fade, wither, droop, shrivel), a reference to the fact that the leaves of this plant droop when touched.

---

13. For details of this practice, see Koopman 1992 and 2002.
14. I have never suspected my neighbours of plotting witchcraft against me.
15. Personal communication, March 2000.

The lexical meaning, then, is 'that which withers, droops, etc.' but a 'deeper' knowledge of the plant's medicinal use will reveal the connotation of this plant as an essential ingredient of a preparation that will cause loss of erection. The associative meanings will be 'impotence', 'infidelity' and 'sexual jealousy'.

Many more examples of these different layers of meaning could be given here, but as later chapters of this book are devoted specifically to examining such meanings in relation to the description of plants, and their medicinal, magical and practical use, no more examples will be given here.

We round off this sub-section with a brief look at two plant names with ambiguous meanings; then explore some levels of meaning in scientific botanical names, and in English, Afrikaans and Sotho vernacular names.

Our first example of a Zulu plant name with an ambiguous meaning is **u(lu)bangalala**. Doke and Vilakazi say this name refers to "Veld plants (different kinds of *Rhynchosia*) used to cause sexual excitement (roots are cooked in milk, and a mouthful taken from time to time)" (1958: 66). This name could be derived from the verb *bangalala* (rage furiously) or from *banga* (cause) + *lala* (sleep, have sex). Either interpretation would suit the way the plant is used.

The second example is **isalanyathi**, given by Boon (2010: 510) as one of three Zulu names for the tree *Ehretia obtusifolia* (Glandular Puzzle-Bush). This name could be derived from the verb *sala* (remain behind) + *inyathi* (buffalo) or it could be from the class 7 prefix *is[i]-* + *ala* (avoid) + *inyathi* (buffalo). Neither of these possible interpretations is recorded by Doke and Vilakazi. My colleague Ndela Ntshangase prefers the second interpretation, but says it is equally likely that the word is a misprint, or a 'mis-remembering' for **ihlalanyathi** (lit., where the buffalo stays).

Some exploration of meaning in botanical names and English, Southern Sotho and Afrikaans vernacular names

Botanical names

As we saw earlier in Chapter 2, when comparing indigenous botanical naming systems with the Latin-based Linnaean system, scientific names for plants may contain one of three types of element:

- a <u>descriptive</u> element that says something about the plant itself, such as *Chlorophytum comosum* (Green Hen-and-Chickens) where *chlorophytum* means 'green plant' and *comosum* means 'tufted with hair' (Pooley 1998: 90). Occasionally this descriptive element is a word from an indigenous language from an area where this plant also grows or originated. For example, *Ocotea bullata* (Black Stinkwood): 'ocotea' is the native name for this tree in French Guiana; and *Caesalpinia bonduc* where 'bonduc' is the Arabic word for hazelnut (Boon 2010: 136) A very unusual descriptive is the generic in *Xymalos monospora* (Lemonwood): Boon (2010: 104) tells us this is an anagram of *Xylosma* (scented wood), the name of a related genus.

- a <u>locational</u> element that indicates where the plant is found, such as *Stachys aethiopica* (Wild Sage) where *aethiopica* indicates "of Ethiopian origin" (Pooley 1998: 182) and *Scleria natalensis* where *natalensis* means "from Natal" (Pooley 1998: 568). Occasionally a locational name refers to the habitat, as in *Gymnosporia nemorosa* where *nemorosa* means 'found in woods and groves'.

- an <u>honorific</u> or <u>commemorative</u> element where either the generic, the specific or both honour a person whose life has been important to botanists and botany. Examples are: *Crinum macowanii* (River Lily) where the specific honours Peter MacOwan (1830–1909, a botanist and teacher who started the South African Botanical Exchange Society) (Pooley 1998: 350); *Stapelia gigantea* (Giant Carrion Flower) where the generic honours the Dutch physician and botanist Johannes von Stapel (d. 1636) (Pooley 1998: 302); and *Strelitzia nicolai* (Natal Wild Banana), a name redolent with European nobility as *Strelitzia* comes from Charlotte von Mecklenburg-Strelitz, who became queen to George III of Britain in 1761, while *nicolai* is named for Tsar Nikolai I of Russia (1796–1855) (Boon 2010: 58, 60).

English names

There is no space for a detailed examination of the underlying meanings of English vernacular plant names, and all I want to do here is to give 20 or so names of trees that relate something of the appearance of the tree or its use in human society.

The first and largest group comprises names descriptive of some physical attribute of the tree:

- Yellowwood (a number of species): a very simple descriptive name that refers to the colour of the timber (cf. blackwood);

- Stinkwood (various spp.): another very basic descriptive, this time a reference to the smell of freshly cut timber. Boon says of the White Stinkwood (*Celtis africana*) "unpleasant smell when cut, hence common name" (2010: 66);
- Silky Needlebush: a descriptive name referring to the fine needle-like leaves of this tree;
- Bead-Bean: the pod-like fruit looks like a string of beads;
- Hairy Caterpillar-Pod: the seed pod resembles caterpillars;
- Itch-Pod: the fruit has irritating hairs;
- Stinking Weed: Boon says the plant "may emit an unpleasant smell" (2010: 144). This tree has the alternative name, Wild Coffee, a reference to the use of the seeds as a coffee substitute;
- Cape or Eared Rattle-Pod: Boon (2010: 152) says the seeds rattle when ripe and gives the Zulu name **ubukheshezane**. Although this name is not in Doke and Vilakazi, they do give the word *ubukheshekheshe* (a rattling);
- Small Knobwood: the name describes the small cone-shaped knobs on the trunk of the tree (see illustration on page 79). Two Zulu names for this tree, **amabelentombi** and **amabelezintshingezi**, reflect the same attribute and both mean 'the breasts of the young maiden(s)'. These are metaphorical names;
- Prickly Thorn: this is about as an overt a name as you can get;
- Horned Thorn: *Acacia grandicornuta*; "*Grandicornuta* = big thorns, Zulu name **umngampondo** means 'like horns'" (Boon 2010: 178). All three names here, the botanical name and the English and Zulu vernacular names, use the same metaphor;
- Spiny Splinter-Bean: Boon (2010: 190) explains the dual nature of this descriptive name – the tree has hooked prickles and the pod splits into one-seeded segments; and
- Red Hook-Berry: another descriptive name with a dual nature. Boon (2010: 102) explains that the tree climbs using woody, hook-like stalks and the fruit has glossy, red segments arranged on a woody hook.

The next three are all metaphorical descriptions:
- Granny Bonnet (Afrik. *moederkappie*): the flower resembles an old-fashioned bonnet;
- Pincushion: the inflorescens of this *Leucospermum* (protea family) looks like a pincushion; and

- Balloon Thorn: in the words of Boon, the spines are "abnormally inflated at base" (2010: 180).

Flatcrown (Afrik. *Platkroon*, *Albizia adianthifolia*) is one of the best-known flat-topped trees. Here is Mark Twain's delightful description of a Flatcrown, written while on a visit to Durban in the late 1890s:[16]

> The "flat-crown" (should be flat-roof) – half a dozen naked branches full of elbows, slant upwards like artificial supports, and fling a roof of delicate foliage out in a horizontal platform as flat as a door; and you look up through this thin floor as through a green cobweb or veil. The branches are japanesich (1989: 646).[17]

Our last five examples of English vernacular names reflect the way trees are used by humans:
- Dragon Tree: the red resin of this tree is sold as 'dragon's blood';
- Shepherd's Tree: Boon explains that the "English name refers to tree in semi-desert areas where it provides the only shade & valuable browse for stock" (2010: 112);
- Fever Tree: Boon (2010: 188) explains that the name 'fever tree' was given by early travellers who associated it with malarial fever. Note also that the bark of this tree is used for fevers;
- Torchwood: the dry kernels of this tree are burnt as torches; and
- Swazi Ordeal Tree: the poisonous bark of this tree is used for trials by ordeal. A person accused of a serious crime is forced to drink water steeped in powdered bark. As Coates Palgrave explains, "If [the accused] vomits he will live and is deemed innocent, but if he fails to do so death is almost certain to follow and he is tacitly assumed to be guilty" (1977: 265).

We turn now to look at underlying meanings in a few Sotho plant names.

---

16. American humorous writer Samuel Clemens (1835–1910), who wrote under the pen name Mark Twain, documented his travels around the world in a number of different books. The description of the Flatcrown comes from his 1897 *Following the Equator: A Journey Around the World,* republished in 1989 by the General Publishing Company, Toronto.
17. This is not a misprint. Twain really does use the word 'japanesich'.

Southern Sotho names[18]

As with my previous caveats about botanical names from Doke and Vilakazi's *Dictionary*, the botanical names given below may have been replaced by later synonyms. We look at only 25 examples of Sotho plant names:

- *kahlana-ea-loti* (*Wurmbea elatior*): *khahla* (plur *li-khahla*) is the generic term for gladiolus or lily,[19] and *le-loti* means 'high mountain' (cf. the Malutis) so this plant is a 'lily-of-the-high-mountains';
- *khongoana-tsa-ngoana* (*Eriospermum cooperi*): *khongoana* is the diminutive of *khomo* (head of cattle), - *tsa*- is a possessive concord and *ngoana* is 'child' so we have here a 'calf-of-the-child';
- *khukhu-e-kholo* (*Eriospermum dissitiflorum*): this name is based on the noun *khukhu* (bulb of a member of the genus *Oxalis*) with the adjective -*holo* (big), which becomes 'kholo' in this construction;
- *lekoto-la-litšoene*: (*Eriospermum bellendini, E. dissitiflorum, Senecio paucifolius* and **S. dieterlenii*): this plant name is based on the noun *le-koto* (edible root, used as medicine) and the plural noun *litšoene* which could refer to a monkey or baboon, a *helichrysum* plant, or a man killed in battle;
- *tsebe-ea-khomo* (*Eriospermum dissitiflorum*): the literal meaning of this name is 'ear of a beast' (head of cattle) so it is the direct equivalent of the Zulu plant name **indlebe-yenkomo**. The leaves of this plant are ear-shaped;
- *leloelenyana-la-lilomo* (**Anthericum humile* and **A. elongatum*): the name is based on *le-loele* (*Kniphofia sarmentosia*) and the plural noun *li-lomo* (precipices, cliffs). The two *Anthericum* species are thus 'little-pokers-of-the-cliff';
- *khoho-ea-lefika* (*Urginea tenella*): the plant name has the literal meaning 'chicken-of-the-rock';
- *khoho-ea-lefika/khoho-ea-mafika* (**Litanthus pusillus*): the same as above with an alternative form 'chicken of the rock<u>s</u>';

---

18. All taken from Pooley (2003) with translation by the writer with the help of A. Mabille and H. Dieterlen's *Southern Sotho-English Dictionary* (1950).
19. Mabille and Dieterlen give *khahla* as "generic term for gladiolus, lily" (1950: 141), but Beaumont (personal communication) points out that *Wurmbea* is, strictly speaking, neither a member of the family *Iridaceae* (where *Gladiolus* belongs) nor of the family *Liliaceae* (where lilies belong).

- *khapumpu-ea-thaba* (*Eucomis bicolor*): the name is based on *khapumpu* (the plant *Eucomis undulata*) qualified by 'of the mountain';
- *metsana-a-manyenyane* (*Ornithogalum graminifolium*): literally 'little-water-which-is-smallest-amount'. This plant is also known as *nko-ea-ntja* (nose-of-dog);
- *lelala-tau-le-leholo* (*Asparagus cooperii*): from *lala* (sleep, lie down) + *tau* (lion) + *holo* (big). In Zulu this name would be **umlalabhubesi-omkhulu** (big where-the-lion-lies). This name also occurs in class 3 as *molala-tau-o-moholo*;
- *tsebe-ea-phofu* (*Haemanthus humilis*): the literal meaning of this name is 'eland's ear'. Pooley (2003: 58) gives the English vernacular name of this plant as Rabbit's Ears and the Afrikaans name *bobbejaanoor* (baboon's ear);
- *lekholela-la-bana-ba-seng* (*Satyrium cristatum*): The noun *lekholela* means 'used up thing, old person', *bana* is 'children' and *seng* appears to be a locative, perhaps a place name; thus 'the dried up old thing of the children of Seng'. It sounds most intriguing. The same noun is the base of *lekholela-la-Matabele* (**Eulophia robusta* i.e., 'the used up old thing of the Matabele');
- *qoqobala-sa-loti* (*Cerastium arabidis*): derived from the verb *qoqobala* (to have only superficial roots) again with the untraceable *loti* (unless, again, this is 'high mountain' – the English name Snow Flower would support this);
- *morarana-oa-mafehlo* (*Clematis brachiata*): *morara* refers to the tattoo made on the arm of a brave man or warrior and *mafehlo* could be related to *mo-fehlo* (dragon-fly). Dragon Fly's Tattoo (if correct) is certainly a splendid name!
- *mohata-metsi-o-monyenyane* (*Crassula natans*) [*natans* = swimming, with floating leaves]: the name is derived from the verb *hata* (step, tread) + *metsi* (water) + *monyenyane* (very little one). The plant is thus a 'little-water-treader'. Cf. the Afrikaans name *watergras* (water-grass);
- *bohobe-ba-setsomi* (*Crassula natalensis*): the literal meaning of this name is 'bread of the hunter', but Pooley makes no mention of this plant being eaten. Also based on a bread metaphor is *bohobe-ba-balisana* (*Crassula nudicaulis*, **C. platyphylla* and **C. turrita*). Here the underlying meaning is 'shepherd's bread'. The only possible reason for this name can be seen in a very brief comment Pooley (2003: 76) makes for another *Crassula* species, namely *C. papillosa* that it is

heavily grazed. The Sotho name for this plant, however, is *serelilenyane* (little *serelile* – a name given to a number of different plant species). *Serelilenyane* is also the name of a number of different species besides *C. papillosa*;

- *setšosa-se-sefubelu* (*Crassula peploides*): from *setsosa* (scarecrow, frightening thing) < *tsoha* (be afraid, fear) + [*se*]*fubelu* (red, dark yellow) i.e., 'The awesome red-yellow thing';

- *noha, noha-ea-loti* (*Epilobium salignum*): as a verb *noha* means 'to divine', as a noun it refers to the plant *Epilobium hirsutum* and various other *Epilobium* species. With the addition *ea-loti*, we have a name meaning 'high-mountain epilobium'; and

- *marama-abaroetsana-a-mamasoeu* (*Sebaea thodeana*): the name is derived from *marama* (cheeks) + *a* (of) + *baroetsana* (maidens).[20] Without further qualification this compound name refers to the plants *Sebaea tranvaalensis* and other *Sebaea* species. When *mama* (which are) and *soeu* (white) are added we get 'White Maidens' Cheeks'.

Although this is of necessity a brief excursion into the semantics of Sotho plant names, it is clear that Sotho vernacular names for plants show very similar underlying meanings to those in Zulu. I have not, in this book, made any analysis of Xhosa and Swazi plant names as they are by and large even more similar to Zulu names in terms of meanings, and are frequently exactly the same name. Let us move on to a few Afrikaans names for plants.

Afrikaans names

Afrikaans names are almost always a translation of English names (or vice versa, of course). I have not counted, but would guess that this is the case for well over 90% of plant species. The following are examples:

- *geelhout* (Yellowwood): *Podocarpus* spp.;
- *stekelsplinterboontjie* (Spiny Splinter-Bean): *Adenopodia spicata* (Boon 2010: 190);
- *kleinknophout* (Small Knobwood): *Zanthoxylum capense* (Boon 2010: 210);
- *moederkappie* (Granny Bonnet): *Disperis fanniniae* (Pooley 2003: 68);

---

20. Mabille and Dieterlen (1950: 332) give the definition of *mo-roetsana* as "girl, maiden, virgin (nubile)".

- *sout-en-peper-blommetjie* (Pepper-and-Salt Flower):[21] *Wurmbea elatior* (Pooley 2003: 48);
- *wildemango* (Wild Mango): *Cordyla africana* (Boon 2010: 150);
- *sandertjiehout* (Sand Peawood): *Craibia zimmermannii* (Boon 2010: 150);
- *appelblaar* (Apple-Leaf): *Philenoptera violacea* (Boon 2010: 164);
- *smalblaarfonteinbos* (Narrow-Leaf Fountain-Bush): *Psoralea glabra* (Boon 2010: 166);
- *Brasiliaanse peperboom* (Brazilian Pepper-Tree): *Schinus terebinthifolia* (Boon 2010: 268); and
- *valslemoentjiedoring* (False Lemon-Thorn): *Cassinopsis tinifolia* (Boon 2010: 318). This one is interesting. At first glance the Afrikaans name looks like an exact translation of the English name. We must remember, though, that *lemoen* is Afrikaans for orange. A lemon is *suurlemoen*.

But English and Afrikaans are not always translations of each other. Consider the following:

The plant *Androcymbium striatum* has the English name Pyjama Flower, but the Afrikaans name *patrysblom* (partridge flower) (Pooley 2003: 48). Another example is *Stachys aethiopica* (Pooley 2003: 78) where the English name is given as Wild Sage and the Afrikaans as *katpisbossie* (cat urine bush). And then again, we have *Alepidea amatymbica*, "named after the Thembu people", with Pooley (2003: 80) giving Giant Alepidea and Larger Tinsel Flowers as alternate English names while the Afrikaans names are *kalmoes* and *slangwortel* (snake root).

The Latin name *Geranium wakkerstroomianum* contains the Afrikaans place name Wakkerstroom (awake stream), yet according to Pooley (2003: 78) there is no Afrikaans name for the flower. Its English name is White Geranium (cf. Boon 2010: 378, who gives the English name Wakkerstroom Red-Pear and the Afrikaans name *wakkerstroomrooipeer* for *Scolopia oreophila*).

By and large the underlying meanings of Afrikaans vernacular names follow much the same pattern as English vernacular names (and no doubt if I were Afrikaans-speaking I would say that the English names follow the pattern of the Afrikaans names!)

---

21. Note the reversal of the elements.

Relationship between meanings in a polysemic entry

We end this chapter on semantics with an interesting example of the relationship between meanings in a polysemic entry. The linguistic term 'polysemy' refers to a word that has a number of different but related meanings. A 'polysemic entry' in a dictionary is one that gives a word with its meanings listed by number. The relationships between the different meanings of a word are often surprising. The single example chosen to illustrate the principle of polysemy also links this chapter on semantics perfectly to Chapter 5 on the uses of *muthi*.

Our starting point is the Caterpillar Bean plant (*Zornia capensis*), which Pooley says was "used traditionally . . . to protect the unborn baby from disease" (1998: 270). She gives the Zulu name **umkhondo**, which intrigued me as I know this word to mean 'track', 'spoor' and 'trail'. Doke and Vilakazi give the following entry for **umkhondo**:

> **umkhondo** . . . 1. track, trail (of passing animals or persons) [with some interesting idioms and examples] . . . 2. disease of new-born babe, entailing a sinking of the fontanelle (believed to be caused by the mother crossing the track of some ill-omened animal, e.g. eland). 3. Species of spreading weed (differing in different localities, e.g. *Zornia tetraphylla*, *Clutia pulchella*) used by pregnant women, tied around the ankle as prophylactic against the *umkhondo* disease (1958: 402).

Doke and Vilakazi's *Zornia tetraphylla* is clearly an earlier synonym of Pooley's *Zornia capensis*, so we are talking about the same plant. The three meanings that Doke and Vilakazi give for **umkhondo** are linked in a most intriguing way. The primary meaning is 'track, trail', but then we go on to what would appear (without the accompanying explanation) to be a completely different meaning: a disease of a new-born baby that causes the "sinking of the fontanelle". The link is the 'folk belief'[22] that such a disease is caused by crossing the track of an "ill-omened animal". Note that there are two parts to this folk belief: namely that crossing the track of an animal can cause a disease in a new-born baby;[23] and that an eland is an animal of ill omen. In order to protect themselves against such problems, pregnant women tie a particular plant around the ankle (the

---

22. I do not like the term 'superstition' as it has decidedly negative connotations.
23. Doke and Vilakazi do not make it clear whether the track crossing that causes the problem takes place before or after the baby is born.

closest part of the body to a track on the ground that may be crossed). Here we have a third folk belief, which in a curiously complex fashion links an unborn baby to the track of an animal and to a particular plant species. It is this third belief that links the third meaning given by Doke and Vilakazi to the second and first meanings.

From this description of the links between animal, plants, diseases of new-born babies and folk beliefs, we now move to Chapter 5 that looks in detail at such links and beliefs.

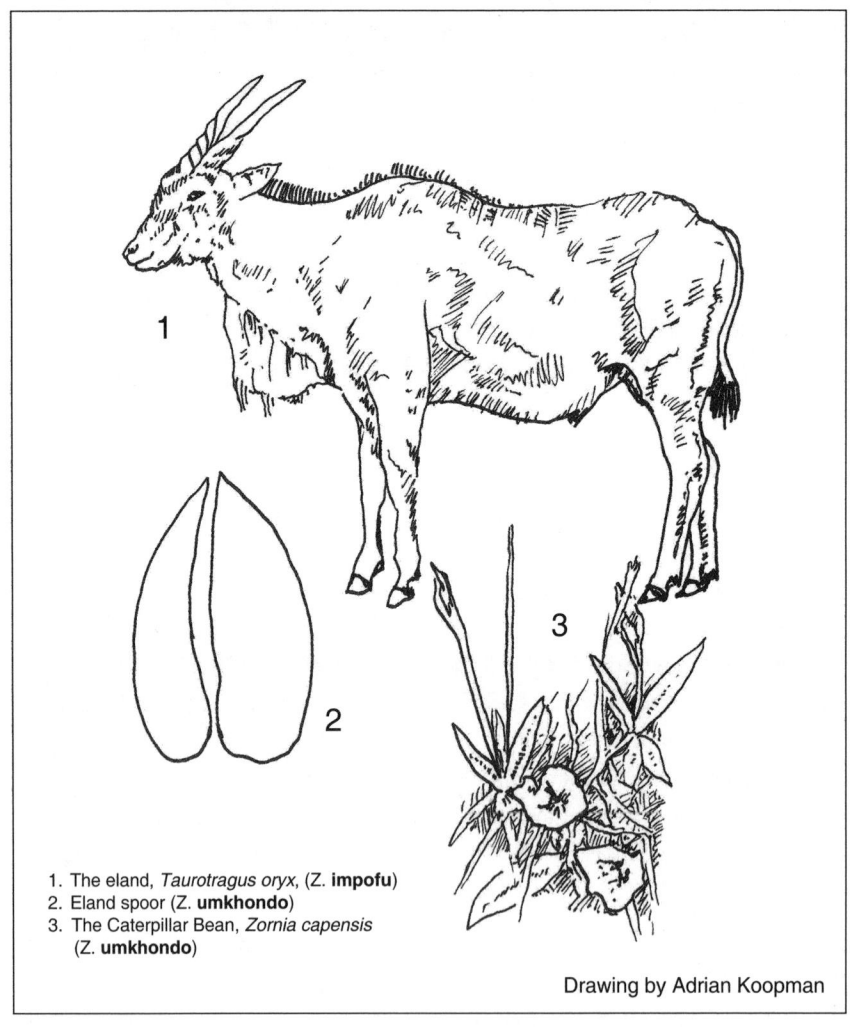

1. The eland, *Taurotragus oryx*, (Z. **impofu**)
2. Eland spoor (Z. **umkhondo**)
3. The Caterpillar Bean, *Zornia capensis*
   (Z. **umkhondo**)

Drawing by Adrian Koopman

5   Using *umuthi*

Introduction

There are baked insects and dried reptiles; the dung of lions in powders
and the fat of the water-sprite in bottles; the shriveled flesh of the white
man and the hardened menses of the baboon; an incongruous assortment
of oddities – Spanish-fly powder, asbestos, glass prisms, washing soda,
flint, spa, crystal, coral, rare geological specimens of every description;
skin and bones of every conceivable animal, and hundreds of barks,
roots, berries and leaves – in a word, choice selections innumerable
and wonderful, medicinal and magical, useful, harmful, and inert, from
the whole range of mineral, vegetable, and animal kingdoms, terrestrial
and marine (Bryant 1970: 19).

"In a word," says Bryant and then goes on to give no less than 25 words. He
could have said 'in a word, *umuthi*' for his list above gives a wonderful picture
of just what *umuthi* is. His 1905 *Zulu-English Dictionary* gives (in addition to
the related meaning of 'tree, shrub') the following:

umu-Ti . . . any substance or preparation which to the native eye
appears to be of a 'medicinal' nature i.e. of vegetable, animal or mineral
extraction, such as ink, blacking, putty, baking-powder, ointment, etc.
(1905: 625).

Berglund's statement on *umuthi* (although he uses the plural form *imithi*) is
also useful:

> A great number of material substances ranging from stones, earth and minerals to practically all the various species of vegetation are believed to contain *amandla*, power. Collectively, these are known as *imithi*, generally translated with the English idiom medicines (1976: 256).

Umuthi does not necessarily, however, refer to material substances for medicinal use only and any material substance used to facilitate a human activity can be called *umuthi*. This could include materials used in the brewing of beer, the smelting of iron, the baking of ceramic pots, the dyeing of grasses and rushes, the tanning of hides, the courting of girls, protection against lightning, and so on and so on. As later chapters of this book deal exclusively with plant names related to these various activities, it is considered important to give a general introduction to the nature of *umuthi*, the different categories of *umuthi*, the properties of *umuthi*, the practitioners who use *umuthi* most extensively, the processes they use, and the knowledge they need to possess in order to use *umuthi* successfully.

This chapter will start with the *umuthi* itself, looking at different categories and properties, then go on to the different practitioners who use *umuthi*, with an overview of the different kinds of knowledge they possess. The focus will then be specifically on medicinal use of *umuthi*, looking at concoction types and the different ways of administering medicine in traditional Zulu society.

The nature of *umuthi*

In the introductory quotation by Bryant, he sums up his list of diverse materials by saying:

> choice selections innumerable and wonderful, medicinal and magical, useful, harmful, and inert, from the whole range of mineral, vegetable, and animal kingdoms, terrestrial and marine.

This seems a useful place to start, and we can note that Bryant has already given us three basic categories of *umuthi*: material from the mineral, vegetable and animal 'kingdoms'.

Let us look at the 'mineral kingdom' first. Bryant lists

> asbestos, glass prisms, washing soda, flint, spa, crystal, coral, rare geological specimens of every description . . .

Details given by Bryant, Berglund, Krige and other writers in a number of books show that mineral *umuthi* also includes soil (such as from the footprints of an intended victim, or where lightning has struck), rocks and stones (of various sizes, especially black and white stones, stones from the seashore, or perhaps stones from the bottom of a waterfall). Stones may be used as they are ground up into a powder and mixed with other materia. Water is also important, especially water from the sea (*amanzi asolwandle*) known to be always cool, and 'living water' (*amanzi aphilayo*) such as that from springs or waterfalls. Bryant talks of "rare geological specimens of every description" and at the Mona Market[1] one can find a wide range of ammonites and other fossils, as well as quartzes, mica, feldspar and so on. As this is a book on Zulu plant names, however, there will be no material in the second half on *umuthi* of mineral origins.

"[S]kin and bones of every conceivable animal," says Bryant, but of course *umuthi* of animal origins also includes teeth, tusks and horn, hair and fur, whiskers, flesh, fat, organs, blood and every other type of animal bodily exudation. Animal origin *umuthi* covers (as Bryant says) both terrestrial and marine, and includes that of insects, birds, snakes and other reptiles, mammals including humans, and even creatures better known in books of fable and myth, such as the *impaka* (witch's cat), *impundulu* (lightning-bird) and *umkhovu* (zombie – resurrected corpse). Although reference will be made in this chapter to various animal origin *imithi*, such will not be covered subsequently.

"[P]ractically all the various species of vegetation," says Berglund, in reference to *umuthi* of vegetable origin and certainly it is remarkable how often the description of a plant found in the Zulu-speaking areas of South Africa (that is, in the source books consulted) has the phrase 'used in traditional medicine' or 'used as *umuthi*'. All parts of plants (trees, shrubs, herbs, grasses, rushes, fungi and others) may be used for their power: bark, roots, leaves, berries, fruits, flowers, seeds, resins and other exudations – there would appear to be no part of a living, growing organism that cannot be used for *umuthi*. Following chapters of this book will give several hundred examples.

---

1. The Mona Market is an open-air market held on every third Thursday of the month on the flat land next to the Mona River, about 15 kilometres north of Nongoma on the Hlabisa Road in northern KwaZulu-Natal. The market is a combined stock, fresh produce, second-hand clothes, indigenous handicrafts and artifacts market, but above all an extremely extensive market for *imithi* of all kinds.

The properties of *umuthi*

"A great number of material substances . . . are believed to contain *amandla*, power," says Berglund. This power, too, can be divided into three sub-categories: chemical, sympathetic and environmental properties.

Chemical properties

As there is a fair amount of published information on the chemical properties of plant materia,[2] this 'mini-section' will concentrate on plant-origin *umuthi*.

Hutchings says of the genus *Eucalyptus*:

> Potential toxins known in the genus include benyaldehyde, borneol, butyraldehyde, butyric acid, careen, carvone, cineole, citral, cuminic alcohol, gallic acid, geraniol, isoprene, isovaleraldeyde [*sic*], isovaleric acid, limonene, linalool, mandelonitrile, phellandrene, pyrocatechol, quercitrin, quinic acid, rutin, saponin, shikimic acid, tannic acid, and valeraldehyde . . . Oil of Eucalyptus is obtained from the fresh leaves of *E. globulus* . . . and contains 70–80% cineole (eucalyptol), α-pinene, phellandrene, terpineol, citranellal, geranyl acetate, eudesmol, eudesmyl acetate, piperitone and volatile aldehydes, principally isovaleric (1996: 219).

Clearly plants can contain a lot of different chemicals.

Hutchings gives the following description of a variety of traditional uses of the Waterberry (Water Myrtle, *Syzygium cordatum*, Z. **umdoni**):

> Bark is used as traditional medicine . . . Unspecified parts are used for respiratory ailments, including tuberculosis, and for stomach complaints and as emetics . . . The Bemba use cold leaf infusions for various stomach ailments, including diarrhea, while poultices of powdered bark, leaves and roots are applied to increase the flow of milk in nursing mothers. Parts of the tree are traditionally worn by pregnant Chewa women as charms against deformities in their babies if their husbands are known to have committed adultery. The Vhavenda use leaves for stomach ailments, colds and fever while bark and roots are

---

2. Especially Hutchings (1996).

used for headaches, amenorrhoea and wounds . . . Powdered bark is used as a fish poison by the Bemba . . . Fruit is acidulous and edible. Leaf infusions have purgative effects. Wood smoke is pleasantly aromatic and used to season milk gourds (1996: 218).

The **umdoni** is certainly a very useful tree in southern African traditional societies. Hutchings gives some indications of the chemical properties of the *Syzygium* genus:

> Potential toxins known in the genus include eugenol, furfural, furfuryl alcohol, methanol, methyl salicylate, naphthalene, valeraldehyde and vanillin . . . Skeels[3] contain a hypoglycaemic principle, the glycocide antimellin and also phytosterin, the alkaloid jambosin, jambolan, essential oil and galli- and ellagi-tannins (1996: 218).

We could continue with several hundred examples of plants with a multitude of traditional uses, not all medicinal, and related lists of their chemical properties. But I am sure the point is made: plants have chemical properties that cause certain effects known to traditional doctors who prescribe them for specific purposes. I am certain that the vast majority of traditional doctors will find the list of chemical terms quoted by Hutchings as baffling as I do. But I am equally certain that the vast majority of doctors will be aware of the physiological effects of the plants they prescribe and that these effects are caused by properties in the plants themselves.

Sympathetic properties

Ndela Ntshangase tells me[4] that in any concoction made by an *inyanga* for the purpose of curing loss of erection in a male, or to <u>cause</u> loss of erection (as when a jealous husband wishes this to happen to his wife's lovers), three ingredients at least are essential. These are **imbune**, *umsundu*, and *intamo yofudu*.

Imbune is the name of the Sensitive Plant (*Mimosa pigra*), a shrub/small tree with leaves that are sensitive to the touch. When touched they droop down, expanding to fullness again after a few minutes. The Zulu name is derived

---

3. Not in *Collins Dictionary of the English Language*.
4. Interview, 6 March 2001.

from the verb *buna* (shrivel up, wither, droop). *Umsundu* is the Zulu word for an earthworm, which, as is well known, also shrivels up or contracts violently when touched and becomes much shorter and smaller. *Intamo yofudu* refers to the neck of a tortoise. The head and neck of a tortoise, when extended from the shell, have much the appearance of an erect penis, but a tortoise, when tapped sharply on the head, will withdraw into its shell.

It is not known whether any of these three ingredients contains chemical properties that would act as aphrodisiacs or anti-aphrodisiacs (i.e., whether they would have any effect on erections or not). What is important is that all three entities <u>behave</u> as in loss of erection. Note that they can be used (with other ingredients) both to <u>cause</u> and to <u>cure</u> loss of erection. The effect would depend on the catalysts mixed with the ingredients: an activating catalyst would <u>cause</u> loss of erection; a negating catalyst would <u>cure</u> loss of erection. The use of such catalysts is discussed in more detail later.

Bryant's *Dictionary* gives a recipe for an *isibethelo* (charm for 'fixing' a girl to you) that requires one of the ingredients to be an *imbambela* (squid, cuttlefish). Again, it is likely that there are no chemical properties in cuttlefish that are aphrodisiac in nature. What is important is that a cuttlefish, like an octopus, has many arms – 'all the better to hold you tight'.

Note that ingredients such as the **imbune** and the *imbambela* are also strong in name with **imbune** meaning literally 'drooper' and *imbambela* meaning literally 'tight holder'. Some writers, for instance Krige, refer to these properties of *umuthi* as 'sympathetic magic':

> In the treatment of many diseases, sympathetic magic plays an important part, and in many case the medicines administered have in reality no curative value at all (1950: 334).

Other writers refer to these properties as 'symbolic associations'. Berglund (1976: 350) compares ingredients used by three different *izinyanga zezulu* (heaven-herds, sky-doctors) making medicines for the control of lightning (again, both to <u>cause</u> and to <u>prevent</u>). Lightning medicines require both black and white ingredients, to reflect the white of lightning and the dark of thunderclouds. In this case the *izinyanga zezulu* were choosing their white ingredients. One chose a 'clear and glistening white ore', the second used milk from a white cow, while the third produced a spark by throwing one stone onto another. All later supported the choices of the others, agreeing that the symbols they had used were suitable. Berglund comments that:

With all three heaven-herds the importance given the symbol by far
outweighed the importance given the materia which carried the symbol
(1976: 352).

He gives a number of examples of 'symbolic associations', distinguishing
between 'antagonistic associations' (like drives away/repels like) and
'sympathetic associations' (like attracts like), of which I list a few here. Four
examples of 'antagonistic associations' are: (1) a heaven-herd used soot and
flint from black stones in medicines for protection <u>against</u> lightning, because
"the colour of the soot and the stone was associated with thunderstorms"; (2)
if dogs sleep too much they are given medicines prepared from *imfinyezi* (a
beetle that curls up when touched) so that they will not be like the beetle; (3)
men who are unable to attain erection are given infusions of **intwalabombo**
(*Rubia cordifolia*), a plant that is soft and bendable (so that they will not be
like the plant); and (4) a man who is considered to be ugly can make himself
attractive by boiling **imbuma** (the bulrush, *Cyperus* sp.) in goat's milk, as
bulrushes are considered ugly and so are female goats with elongated udders
(Berglund 1976: 353, 354).

Examples of 'sympathetic associations' (where like causes like) are: bald
men treated with medicines from gardens with a profuse growth; talkativeness
treated with parts from a sheep, considered a very quiet animal; and a man with
very short legs treated with bones from a stork. Red is obviously the colour of
blood, so any plants that exude red sap will be used in concoctions designed
to treat blood problems, menstrual problems, and problems associated with
childbirth. Similarly, *abathakathi* (witches) will use plants and other material
that exudes red liquid in order to <u>cause</u> heavy bleeding in a victim (Berglund,
1976: 355, 356). Note the following entry from Bryant's *Dictionary*:

> **u-Mopo** (< *opha* – bleed) Shedding of blood . . . certain forest tree
> which when cut emits a reddish sap; certain sea-animal (perhaps a
> species of anemone) which when cut emits a reddish fluid like blood
> and is much sought after by Native doctors for purposes of *takata*
> (1905: 301).

Berglund also refers to 'visual symbolism', giving examples of a herbalist who
prepared love medicines for girls without lovers from scrapings collected from
the fork of trees "because it is like a woman", and the treatment of women with
underdeveloped mammary glands with medicines made from pawpaw fruit
because "the paw-paw is like the breast. Even the nipple is there" (1976: 357).

Environmental properties

There is little in the literature that stresses the significance of the environment of the harvested *umuthi*, but my colleague Ndela Ntshangase has on a number of occasions stressed how important it is that certain medicines, desired for specific effects, are harvested from a high, windy place, in a deep, shady forest, or (very commonly) from near the sea where the salty spray contributes considerably to the power of various materia. Berglund has stressed how important it is that rain-making medicines include materia collected from in or near the ocean:

> *Ukhuningomile*, sea-weed, fetched from the rocks in the ocean, form an essential part of the [rain-making] medicine and without it the medicine is of no avail. Equally important is *ikwindi*, mussel shell-fish, and *isikhukhukhu* sea urchin, without which the medicine is not efficient (1976: 56).

Even the environment on its own (i.e., not as a source of materia) is important. Berglund quotes various rain makers who say that from time to time they must go to the ocean and simply "sit, staring out over the sea for considerable lengths of time" (1976: 57).

In connection with this, Ndela Ntshangase notes that re-planting certain trees in nurseries[5], plantations and other places is no good. The natural habitat is necessary; the *muti* power comes from the environment. For example, bark from the Black Stinkwood (*Ocotea bullata*) must be from thick indigenous forest (*ihlanze*); *umuthi* from the same tree, when found in open country (*inkangala*) is generally not as powerful as that from dense forest. Under certain circumstances, however, 'open country stinkwood' is essential.

This emphasis on the environment is frequently reflected in closely related species of the same genera (or considered similar in Zulu thought patterns – by no means the same thing!) having the same names, distinguished only by a locational epithet. See, for example:

- **isibangamlotha**: Tassel Berry (*Antidesma venosum*);
- **isibangamlotha-sehlathi** (forest **isibangamlotha**): Warty Whiteberry-Bush (*Flueggea verrucosa*);
- **umthunzi**: Coast Red Milkwood (*Mimusops caffra*); and

---

5. He is referring to the practice of some conservation bodies to plant certain over-utilised plant species in nurseries in order to relieve pressure on the same species in the wild.

- **umthunzi-wehlathi** (Forest **umthunzi**): Forest Red Milkwood (*Mimusops obovata*).

As we saw in Chapter 3, the plant name **ihlamvu** is particularly extended by epithets referring to specific habitats:
- **ihlamvu**: *Gloriosa superba, Callilepis laureola* and others;
- **ihlamvu lasenhla** (down-country **ihlamvu**): *Sandersonia aurantiaca*;
- **ihlamvu lasolwandle** (coastal **ihlamvu**): *Gloriosa superba*;
- **ihlamvu lehlathi** (forest **ihlamvu**): *Littonia modesta*; and
- **ihlamvu elimpofu lasenkangala** (pale open-plain **ihlamvu**): *Disa stachyoides*.

The users of *umuthi*

There are five groups of people I would like to consider here:
1) the *inyanga* (herbalist, traditional doctor, medicine man);
2) the *isangoma* (diviner, shaman);
3) the *inyanga yezulu* ('sky-doctor', 'heaven-herd');
4) the *umthakathi* (witch); and
5) the *umuntukazana* (the common man, man-in-the-street, householder).

The inyanga

The *inyanga* is unquestionably the greatest user of *umuthi* in traditional Zulu society and also has the greatest knowledge. Anyone can become an *inyanga*, but in practice the *inyanga* is usually a male and the son, grandson or nephew of an *inyanga*. An *inyanga* starts young, helping his father (grandfather, uncle) harvest and prepare *umuthi*, and so spends a number of years as an apprentice learning as much as he can before branching out into private practice later in life.

The term *inyanga* is used in Zulu much the same way as the term 'doctor' is used in English. In English, one assumes that someone addressed as 'doctor' is a medical doctor, until specifically told that the person is a Doctor of Philosophy, or a Doctor of Music, Law or Literature. It may sometimes be necessary to qualify the term as 'Doctor of Medicine'. In the Zulu language, the term *inyanga* means 'person with great skill and knowledge in a certain area', but normally it is assumed that an *inyanga* is a traditional doctor of medicines. One can also talk of *inyanga yensimbi* (lit. doctor of iron – a celebrated blacksmith), *inyanga yokubumba* (lit. doctor of moulding – a highly skilled maker of clay pots), or *inyanga yengoma* (doctor of song or music – a highly skilled musician). It may sometimes be necessary to qualify the term *inyanga* as *inyanga yemithi* (doctor of medicines) or *inyanga yokwelapha* (doctor of healing).

As the term *inyanga yokwelapha* (doctor of healing) indicates, *izinyanga* spend most of their time prescribing medicines for healing illnesses and disease. But *izinyanga* do also prescribe and provide medicines for helping you win your court case, have success at the races, find a job, keep a job, make yourself attractive to girls, keep your girlfriend from straying, punish your wife's lovers, and even, in today's competitive sporting world, ensure that <u>your</u> team performs well, while simultaneously ensuring that the opponents perform badly.

The indigenous knowledge of an inyanga

A properly trained and qualified *inyanga* is many things: he is a botanist, a chemist, a pharmacist, a diagnostician and a physician.

As a botanist, he needs to be able to identify the plants he needs, to know where to find them, to know <u>when</u> (what season, what time of day) and <u>how</u> to harvest them, what parts to use, and how to conserve the population of each species for further use.

As a chemist he needs to be familiar with the properties of the plants he collects: which will be suitable as soporifics, which as vermicides or anthelmintics, which as aphrodisiacs, which as purgatives. He needs to know which are poisonous, which will stimulate the heart, which will stop bleeding, and which will ease cramps, chest complaints or urinary infections. He needs to combine knowledge of chemistry and botany to know which parts of the plant contain the active ingredients needed: the roots? – the resin? – the seeds? – the bark?

As a pharmacist, the *inyanga* needs to know how to prepare concoctions. What to mix with what, and in what quantities. Does the plant or other material need to be dried, crushed, ground or soaked in oil? More details of this knowledge and skill will be evident later under the separate heading 'Concoctions'.

An *inyanga* needs to be a diagnostician. When a patient presents with something obvious like an axe wound or snakebite, this is one thing. It is quite another to be able to determine poor circulation, liver complaints, infertility, heart murmurs and a variety of inner infections.

And then finally, a qualified *inyanga* is a physician. He combines all the skills mentioned above to decide which preparations of which *umuthi* will cure particular problems. He treats, he heals. As with any other doctor, he may refer a patient to a specialist: either another *inyanga* who has a particular reputation for say, infertility, or heart problems, or to an *isangoma* if he suspects that a problem may be caused by angry *amadlozi*,[6] *abathakathi*, jealous neighbours, and so on.

Administrative procedures

As a doctor, he will be familiar with all the following ways of administering medicinal 'muti':

- *ukugwinya* (to swallow): oral ingestion of
  - (a) *isichonco*: a cold infusion of various prepared medicines, or
  - (b) *imfudumezelo* (< *fudumala* (become hot, warm)) > *fudumeza* (make hot, warm)): a hot infusion, or
  - (c) *impeko* (< *pheka* (cook)): a simmered decoction;
- *ukukhwifa*: to blow out liquid in a fine spray. The *inyanga* (or patient) takes a mouthful of a liquid concoction and blows it out over the affected part;
- *ukugcoba*: to rub medicine into an open wound;
- *ukugcaba*: to make incisions in the skin and rub in medicine;
- *ukumunca* (to suck): to apply a poultice to an inflammation so that it draws out the 'poison';
- *ukugquma*: to seat the patient with body or head in steam, vapour or smoke. A variation of this is to get the patient to inhale the smoke from *umuthi* burnt on a potsherd;
- *ukuphalaza/ukuhlanza*: to use emetics for 'deep-cleansing';
- *ukutshopa*: to 'prod' medicines into the skin with sharp implements (such as thorns, porcupine quills)[7]; and
- *ukuncinda*: to suck medicines from the tips of the fingers.

Bryant's *Zulu Medicine and Medicine-Men* (1970) gives a number of examples of how *izinyanga* actually mix *imithi* and treat patients. The following are just a few examples:

---

6. *Idlozi*, plural *amadlozi*, variously referred to in the literature as the ancestors, ancestral spirits, the 'living dead' and shades.
7. This is as close as traditional Zulu medicinal practices get to the Euro-Western medical practice of injecting medicines.

the roots of the **uMampeshana** (**Oldenlandia decumbens*) herb, crushed, infused in boiling water and drunk . . . are used for . . . palpitation (1970: 67).

For heart problems:

Certain species of **inTsema** or dwarf euphorbia (*E. pugniformis* and *E. bupleurifolia*) are . . . employed, the roots being dried, burnt, and the ashes rubbed into incisions made about the affected parts (1970: 67).

For rheumatism:

the roots of the **uShaqa** (*Berkheya* sp.) may be boiled, its leaves pounded and mixed with cold water, which is then added to the boiled root-decoction, and the mixture used to foment the painful limb (1970: 66).

For syphilis, some *izinyanga*

administer internally a decoction of the roots and leaves of the **uNjalwana** veld-herb, also a decoction of the leaves of the common aloe or **umHlaba** (*A. ferox*); and they sprinkle on the external sores the same leaves charred and ground, or better, a paste of the bruised leaves of the **uZipho** or **iKhambhi leziduli** (*Cardiospermum halicacabum*), the **umDlonzo** (*Mikania capensis*), the **umSintsi** (*Erythrina caffra*), and the **uXhaphozi** (**Ranunculus pinnatus*), is laid on as a poultice (1970: 59).

The **isangoma**

The *isangoma* (plural *izangoma*) has a number of functions in traditional Zulu society. She[8] functions primarily as a shaman, that is to say as liaison between the worlds of the living and the supernatural. Her competency lies mainly in spiritual matters. When people have misfortune, she is able to determine whether this is 'the wrath of the ancestors', witchcraft or the evil machinations resulting from envy and jealousy.

---

8. The *isangoma* may be of either sex, but in practice most *izangoma* are female so the feminine pronoun is used here.

The *isangoma* interprets for the *amadlozi*, 'smells out' witches, finds out the reason for dreams, helps find lost property, and identifies disturbed minds as either being mental sickness, psychological stress, or (very often) a sign that the ancestors have chosen a particular person to become an *isangoma*. Many *izangoma* are specialists: some are renowned for their interpretation of dreams; others as identifiers of witches (in which case they are often known as *izanusi*, lit. smellers-out); yet others highly regarded as trainers of young initiates into the 'mysteries' of divination.

Izangoma have much knowledge of *imithi*, not nearly as much as the *inyanga*, and not especially in the field of physical healing, but more especially of medicines that help troubled minds, calm hysteria and allow communication with the spirit world.

One plant particularly associated with the *isangoma* is **impepho** (the Everlasting Flower). Berglund (1976: 113) gives this as *Helichrysum miconiaefolium* [= *miconiifolium*], but Pooley (2003) records **impepho** as the Zulu name for *H. natalitium* (102), *H. odoratissimum* (146) and *H. cymosum* (148). The Helichrysum species are burnt in a potsherd and the smoke is believed to attract the *amadlozi*. One of Berglund's diviner informants explained to him that the Everlasting Flower is similar to the shades (*amadlozi*) in that neither withers away, "so they are friends (*abazalwane ngokufanana*)" (1976: 113).

The **inyanga yezulu**[9]

The term literally means 'doctor of the heavens' (*izulu*: sky, weather, heavens)[10] and refers to a person who is able to control the weather. Berglund (1976) distinguishes between 'rain-makers' and those who control thunder and

---

9. For considerably more detail on the role of the *inyanga yezulu*, especially in the control of lightning, see Koopman (2011).
10. The word has come to refer to one of the dominant black groupings of south-eastern Africa because the original progenitor took his name from this word. The stem *-zulu* occurs in a number of forms: *izulu* (sky, weather), *uZulu* (a person whose personal name is 'Zulu', the Zulu nation as a single entity), *oZulu* (a person called Zulu and his friends; a group of people all using the name Zulu as a personal name), *abakwaZulu* (those of the house of Zulu: the clan using Zulu as a clan name), *amaZulu* (members of the modern Zulu nation), *kwaZulu* (the place where the members of the Zulu nation live, now prefixed to 'Natal' to form the province name KwaZulu-Natal), *isiZulu* (the language spoken by the Zulu nation) and *ubuZulu* (the culture and heritage of the Zulu nation).

lightning, and only refers to the latter as *izinyanga zezulu* ('heaven-herds'). *Izinyanga zezulu* only use *umuthi* for the purpose of the control of weather. Two major functions are to bring rain in the case of drought and offer protection against lightning strikes. Lightning strikes are a major cause of death in rural KwaZulu-Natal and as it is strongly believed that most are the result of malicious intent, it is extremely important to insure against them by employing the services of a competent *inyanga yezulu*. (Having said this, it is the same *inyanga yezulu* who may be hired and paid to cause lightning strikes!)

Berglund gives details of steps taken when thunderclouds gather. Anything remotely white, shiny or light in colour is removed, lest it attract lightning:

> Water, looked upon as white, is thrown out at some distance from the household. All shiny metal tools and containers are brought indoors. Cattle which are white or have much white skin are removed from the cattle-enclosure and driven away . . . Mirrors are stored away . . . every form of fire is extinguished (1976: 52).

I recall that in the 1970s a commercial firm in Natal was manufacturing solar reflectors, a device with a shallow bowl-shaped mirror that reflected and concentrated the rays of the sun onto cooking pots, which would have cut down on the need for firewood enormously. They were finding it impossible to sell these in rural KwaZulu-Natal and in many cases were not even permitted to demonstrate their use. They called in a social anthropologist from the University of Natal to explore this reluctance to use this cheap and inexhaustible solar power and, as can be expected, it was quickly traced to the thought patterns mentioned above.

The umthakathi

The *umthakathi* (witch, plural *abathakathi*) is hated and feared in Zulu society. He or she[11] is dedicated to the destruction of harmony, fertility, productivity, and anything else that constitutes social norms and goals. If a marriage is a happy one, a nearby *umthakathi* will try to destroy it. If a maize crop is healthy, the *umthakathi* will send a blight. If a man's cattle are healthy and numerous, it is in his interests to protect them against *abathakathi*. Almost every *umuzi*

---

11. There is no clear domination of one sex or the other.

(homestead) will regularly have *imbozisa*-type plants (see below) planted around the edges to deter the actions of *abathakathi* and will as regularly employ an *inyanga* to sprinkle (*chela*) *izintelezi* ('blocking' medicines) around the homestead.

It is widely believed (even in the twenty-first century) that *abathakathi* employ familiars, which include *imikhovu* (zombies, resurrected corpses; singular *umkhovu*), *izimpaka* (a witch's cat) and *izimfene* (baboons). These familiars have two main tasks: to collect *insila* (lit. body-dirt or body-essence such as fingernails, hair or excreta of an intended victim); and to deposit *umeqo* (see below for an explanation) and other malicious preparations where they will come into contact with intended victims.

Abathakathi specialise in *imithi* that have poisonous and other deleterious properties. Bryant gives an extensive list of plants with 'dangerous' properties (1970: 20). These include: **ilozane** (*Tephrosia macropoda*), **imfulwa** (**Ophiocaulon gummifera*), **umahedeni** (**Phytolacca abyssinica*), **inhlungunyembezi** (*Acokanthera thunbergii*) and **umahlabekufeni** (*Croton gratissimus*). Chief amongst the 'dreaded' plants must be the **umdlebe** (*Synadenium cupulare*). Boon says of this tree:

> *Synadenium cupulare* – Dead-mans Tree; dooiemansboom . . . sap very poisonous, causes blisters, headaches & nausea. Despite being feared because it allegedly lures people to their deaths, leaves & stems used medicinally . . . (2010: 258).

Ndela Ntshangase says[12] that the **umdlebe** tree senses blood in humans and animals, stretching out to catch them and sucking them dry after injecting them with poison. Coates Palgrave adds even more lurid detail:

> Africans are sure this is one of the most evil of trees: they firmly believe that it cries like a plumed snake (which is said to bleat like a goat!) and that it lures people and animals towards it in order to kill them. They are convinced that the ground around the trees is white with the bones of slain animals and that birds flying above will fall dead from the sky (1977: 454).

---

12. Personal communication.

Another plant species that would appear to lend itself to use by *abathakathi* is the Dune Poison-Bush (*Acokanthera oblongifolia*). Coates Palgrave explains why:

> The fruit is said to be highly toxic, especially when unripe. Bushmen dip their arrow-heads into the potent poison yielded by the wood and bark and this has on occasion been put to use by suicides and would-be murderers (1977: 781).

The umuntukazana

By this word I mean the non-specialist, the ordinary householder. Any member of Zulu society has basic knowledge of a few medicinal herb and protective charms. Most will keep or know where to find, plant and other materials to help a bad cough, disinfect a superficial wound or bring down a temperature. Most people know what plants to plant near and around the homestead to provide at least some minimal protection against witches, bad luck and general malice. These 'household remedies' are the equivalent of the medicine cabinet in a Euro-Western household, with its Disprin, Dettol and plasters. These household remedies are commonly known as *amakhambi* (see under the heading 'Named medicine types' in this chapter for an explanation).

Bryant lists three species of plant of which "every Native [*sic*] fortunate enough to procure them habitually carries about with him a supply of one of these drugs" (1970: 33), chewing a portion of the root in cases of indigestion and heartburn. These are **umondi** (**Chlorocodon whitei*), **indawo** (*Cyperus esculentus*) and **umhlwazi** (*Catha edulis*). Of this last, Boon says "widely used & known as 'khat' in trop Afr & Arabic states for the habit-forming stimulant cathinone found in the leaves (brewed as tea or chewed)" (2010: 290). The Swazi name *umlomo-mnandzi* (sweet mouth) clearly relates to the use mentioned by Bryant, namely "as a cure for foul breath" (1970: 34).

Concoction types

Ordinary householders may use just one type of *umuthi* at a time, but almost all medicines, charms and 'spells' are the result of a combination of different materia. *Izinyanga* and other expert *umuthi* users are famously reluctant to divulge their professional knowledge about exactly what is combined, in what amounts and in what manner, and there are very few examples in the literature.

The following two, however, are examples of preparations/concoctions, and both are from Bryant's *Dictionary*:

> if the *umzanyana womfazi* (the placenta of a woman) and the *umhlapo wehashi* (that of a horse) be mix[ed] together with *idhlaligwavuma* (human fat) and *umdhlebe* (a poisonous bush) and *umopo* (a certain sea-animal) and *ifelakhona* (a certain mollusc ['probably a limpet' – page 142]) and one or two other ingredients, a powerful *umbulelo*[13] will be prepared! (1905: 55).

> **isi-Betelelo**. Any love-medicine used for the purposes of ['fixing' a girl]. N.B. Take of the *i-mBambela* (cuttle-fish), *u-Manaye* (plant), *u-Nginakile* (plant), *u-Zililo* (plant), *ama-Futa engwe* (leopard's fat) and *u-Lukuningomile* (plant), each a part and mix with the spittle of any particular girl and your own; place all, carefully covered up, beneath a projecting rock in some precipice, and the girl is 'fixed' firmly to you against all comers! (1905: 33).

Berglund lists ingredients that must be added together to make rain-making medicine. These include *ukhuningomile* (seaweed), *ikwindi* (mussels), *isikhukhukhu* (sea-urchin), "pieces of vegetation cast up by the waves on the shore before they have dried", fat from a whale, and well-ground pebbles, both black and white, that were picked up along the sea shore (1976: 56).

Berglund also gives (1976: 49) the ingredients of medicines for driving off storms. The most important ingredient is the fat of the lightning bird (*inyoni yezulu*). To this should be added the fat of *umonya* (*Python sebae*, the Natal rock snake), the *imbulu* (iguana, *Varanus albigolaris*) and the fat of a <u>black</u> sheep. The fat and the skin of *imbila* (*Hyrax capensis*, the rock rabbit) and the skin of an otter (*umanzi*) are added as well. The flesh of a tortoise and the feathers of a peacock are also required. The last two ingredients are botanical in nature: the crushed stem of **umunka** (*Maerua angolensis*) and the crushed bulb of the red Ifafa Lily (*Cyrtanthus stenenthus* [= *stenanthus*], **impingizane encane ebomvu**). It is apparently important that the total number of ingredients is ten as there are ten fingers on two hands.

---

13. See explanation of the term *umbhulelo* in the next sub-section 'Named medicine types'. Bryant uses the older (pre-1950) spelling 'um-Bulelo'.

Catalysts

It has already been mentioned a number of times in this chapter that the same medicines may have opposite effects. Ndela Ntshangase has explained[14] that these contrasting effects will depend on what 'catalysts' are used. As he explains it, in any preparation there are a number of different ingredients and a catalyst, which, according to the effect wanted, may be either a 'starter' (activator) or a 'stopper' (neutraliser). So, for example, if a person requiring medical help with erection problems goes to an *inyanga*, that doctor will combine medicines with aphrodisiac properties and other medicines with symbolic properties (neck of tortoise, bulbs or roots that suggest male genitalia, etc.) and then add an 'activator'. On the other hand, a man who requests an *inyanga* to help him 'de-activate' the lovers he is sure his wife is entertaining (through medicines given to the wife) will be given effectively the same combination, but this time with a 'neutraliser', or 'de-activator'.

The details given by Berglund (1976) on the preparations made by *izinyanga zezulu* for lightning control show much the same thinking. The heaven-herd is usually asked to provide medicines for protecting a household or individual against lightning, but may also be asked, usually by someone envious of another's wealth or good fortune, to cause lightning to strike that person. Again, in each of these contrasting intentions, the basic mixture is likely to be the same, but for protection 'lightning-powerful' *umuthi* will have a neutraliser added, while to cause lightning to strike in a particular place, an activator will be added.

Here is a final cautionary note on 'sympathetic properties' from Ndela Ntshangase:

> When mixing ingredients for a particular purpose, ingredients with 'sympathetic' properties are mixed with those that have 'chemical' properties. But the person doing the mixing must watch out for *ukudlana* (lit. eating one another – when ingredients cancel each other out) and *ukwelekana* (inter-domination – when one ingredient dominates too much over another). Some *imithi* 'eat' other *imithi* when mixed together. A good, trained *inyanga* knows when, what and how much to mix together, and also when not to mix.

---

14. In personal communications and in seminars.

He gives the following example:

> The plant **isidondwane**[15] because it is *intelezi* has 'blocking power',
> i.e., it is a neutralizing agent – it turns negative forces into something
> with no power. *Isidondwane* can also be added to concoctions to
> make neutralisers: chop and add to other neutralizers; and add some
> **umalala**[16] if you are having a party/function, e.g., *umshado* (wedding),
> as this will stop people becoming aggressive and fighting. Any such a
> concoction is an *intelezi*.

Named medicine types

Certain medicines and preparations fall into named categories if they constitute
a group with a common purpose. Under this heading we can consider the natures
and uses of *amakhambi, isibethelo, umbhulelo, imbiza, impepho, amakhubalo,
imfingo, umkhando, intelezi* and *imbozisa*.

The following details are based on definitions and explanations given by
Bryant in his *Dictionary* (1905):

- *intelezi*: this is a general name for all those medicinal charms whose
  object is to counteract evil, such as when a traditional doctor neutralises
  the poison of an *umthakathi*, or the flashing of lightning. *Intelezi* is
  generally administered by the process known as *ukuchela* (to sprinkle),
  and not carried about on the person as is the case with *imfingo*;
- *imfingo*: another name for any medicine used to neutralise evil. It is
  carried on the person as Bryant illustrates in the following: "*Ngimpatele
  imfingo*, I am carrying an *imfingo*-charm for him, *i.e.* against him – as
  every man was accustomed to do when going out to face any enemy or
  danger" (1905: 146);
- *imbozisa*: this noun is derived from the verb *bozisa* (cause to rot),
  derived in turn from *bola* (rot, decay). It is used of any medicine or
  plant employed to cause decay or 'dying-off' in people, crops, etc.,
  such as that used by an *umthakathi*. An *imbozisa* may also function as
  a counter-remedy or antidote to nullify the effects of another *imbozisa*;

---

15. Boon (2010: 98) **isidondwane** = *Portulacaria afra, spekboom*.
16. Pooley (1998) = *Jasminum angulare*, Wild Jasmine (162); = *Osyridicarpos
    schimperianus* = **umayime** "used as a protective charm" (250).

- *umbhulelo*: this refers to any type of poison or injurious medicine placed in a homestead or along paths by an *umthakathi* for the purpose of causing fatal disease in those who should come in contact with them;
- *umqotho*: this is any powdered medicine mixed with snuff with the object of killing a person. It is derived from the verb *qotha*, with the meanings 'grind dry', 'make smooth', and 'make an end of thoroughly, finish off entirely';
- *idlaligwavuma*: this word, literally 'one who growls while eating', refers to human fat used by an *umthakathi*. Another word for human fat used for evil purposes is *iphumalimi* (one who comes out of his hut standing erect, a reference to a white person who does not have to kneel down to get in a door, as would be the case for a Zulu person entering the low door of a grass hut). Compare these two words to *isothamlilo* (one who basks by the fire), a reference to a black person, as well as to the fat of a black person as used by *abathakathi*;
- *umkhando*: this word is derived from the verb *khanda*, which in addition to meaning 'pound, beat, hammer', 'work metal or stone' and 'repair' also means 'doctor with charms or medicines'. The word is used of various kinds of medicinal charms, mostly stones, quartz, etc., but sometimes roots – used by traditional doctors to gain an influence over others and take away their power. Bryant says the word is also used of a "large round stone formerly used by Native blacksmiths in place of [a] hammer" (1905: 292);
- *isibethelelo*: derived from the verb *bethelela* (prepare a charm to ensure for one the affections of some girl), this word refers to any medicine used for the purposes of securing the love of a particular girl. Doke and Vilakazi (1958: 75) add that the word is also used of any medicinal protective charms such as sticks or pegs used to protect against thunderstorms, hail and evil influences;
- *ikhubalo*: Bryant defines this word as "Any Native wood-medicine (which is kept or sold in the lump), as medicinal roots, bark, and the like, not leaves, bulbs, stones or animal powders" (1905: 324). He goes on to note: "N.B. *Amakubalo* are eaten always upon the death of one of the family, in order to strengthen against ill-effects that might otherwise follow. They must be eaten before food is taken. Thus, should a person die in a kraal, *amakubalo* must be eaten by all the remaining inmates thereof, except in the case of a wife, when only the children of her hut take the medicine, she being eaten for, of course, by all the

members of her own parental kraal. Cp. *um-Lawu*". Doke and Vilakazi give the following words derived from *ikhubalo*: "(1) *ikhubalozimbe* (< *ikhubalo* + *zi* (itself) + *mba* (dig), i.e., medicine that digs itself): species of unidentified medicinal tuber which has no surface leafage; (2) *inkubalwane* (< *ikhubalo* + diminutive *-ana*): 1. Small medicinal dose; 2. Herbalist's name for the forest tree *Rapanea melanophloeos*, generally called *isiqalaba sehlathi*" (1958: 408);

- *iqhuzu*: another word for *ikhubalo*;
- *ikhambi*: this word refers to any common medicinal herb used as a common 'household remedy'. It is not used of 'professional' medicines, such as those used by traditional doctors;
- *idoyi*: Bryant explains this as "Medicine of any kind taken by the members of a family immediately after the death of one of their number and previous to taking any food, in order to 'brace up' (*qinisa*) their bodies" (1905: 171);
- *umlawe*: this refers to any strengthening or purifying medicine taken by a man who has buried another, or who has had contact with the stock in a cattle byre in which a death has occurred;
- *umkhumiso* (< *khuma* (take powdered medicine) + the causative suffix *-isa*): This refers to any powdered medicine, or, more specifically "Roots, bark, etc., mixed with powdered waxbills [small birds] . . . and administered to cattle, goats, etc., as a tonic or condition-powder" (Bryant, 1905: 330);
- *ubulawu*: love potion or other medicinal charm of any kind, commonly used by young men in their dealings with girls. The word is used in the expression *ukuthela ngobulawu emehlweni* (to throw dust in the eyes);
- *ikhathazo*: this word is used of any small veld herb, the roots of which are worn around the neck and occasionally nibbled against an *umkhuhlane* (cold, fever). Doke and Vilakazi specifically identify this word as referring to the umbelliferous plant *Alepidea amatymbica* "once common in the mistbelt areas: roots used for colds and influenza" (1958: 384). The word is used in the saying *Ikhathazo labuyela emfuthweni* (lit. the medicinal plant returned to the doctor's bag; i.e., the treatment was unsuccessful). In the plural form *amakhathazo* this word refers to general household and domestic remedies; and
- *umeqo*: derived from *eqa* (jump over, step over); an *umeqo* is a harmful concoction buried in the soil in a path frequently used by the intended victim.

Conclusion

In this chapter I have tried to give a general background to the Zulu traditional cultural use of *umuthi*. We have looked generally at the nature of *umuthi*, briefly at the different users of *umuthi* and at different ways of concocting and of administering *umuthi*, and have ended by looking at different broader categories of *umuthi* depending on their aim or target. The focus in this chapter has not been specifically on plants, or on plant names. In Chapter 7 we look specifically at plant names that refer to their medicinal use. There will inevitably be a little overlap.

This chapter and the four that precede it have been intended to give a background to the chapters that now follow and deal specifically with Zulu plant names.

6 Names referring to the plant itself

Introduction

For the sake of convenience, we can divide the names in this chapter into three categories:

- those that describe the plant itself and various of its physical characteristics. An example in this category is **inkalamasane**, the Zulu name of *Euphorbia natalensis*. The Zulu name is derived from the verb *khala* (weep) and the diminutive form of the noun *amasi* (sour milk), so the literal meaning of the Zulu name is 'that which weeps milky tears', a reference to the milky latex of this euphorbia;

- those that describe where the plant can be found, i.e., its characteristic habitat. As an example let us take **umamentabeni**, the Zulu name for the Cork Bush (*Mundulea sericea*) found on rocky hillsides. The Zulu name consists of the noun class prefix *u-* + *ma* (characteristic) + *ma* (stand) + *entabeni* (on the hill). The literal meaning of the name is thus 'that which characteristically stands on a hill';

- those that describe the relationship between the plant and the animal world. For example, both the Sand Canary-Berry (*Suregada zanzibariensis*) and the Zulu Loquat (*Oxyanthus latifolius*) are known by the Zulu name **umdlankawu** (literally, what the monkey eats, from *dla* (eat) and *inkawu* (monkey)) (see Plate 20).

In Chapter 2, in a brief description of the semantic features of Latin-based scientific botanical nomenclature, I mentioned that in terms of underlying meanings we could divide both generic and specific names into three categories: (a) those describing a physical characteristic of the plant; (b) those referring to the place where the plant is characteristically found; and (c) those commemorating the name of a person. Let us look briefly at examples of these three categories again to see if they overlap in any way with the three categories of Zulu plant names I have mentioned above.[1]

(a) Physical characteristics of the plant: the Sickle Bush is referred to by botanists as *Dichrostachys cinerea*, where *Dichrostachys* (from the Greek *di* (two) + *chroos* (colour) + *stachys* (spike)) refers to the flowers that hang in spikes, yellow above, mauve to pink below (Boon 2010: 196) and *cinerea* means 'ash-coloured' (reference not clear, but possibly to the bark). The Splendid Witch-Hazel goes by the Latin moniker *Trichocladus grandiflorus*, where *Trichocladus* means 'with hairy branches' and *grandiflorus* means 'large flower' (Boon 2010: 124, 126). There is a considerable amount of overlap between this category and the first of the three Zulu plant name categories mentioned above; i.e., names that describe the plant itself and aspects of its physical characteristics.

(b) Places where the plant is characteristically found: in the tree names *Cyathea capensis*, *Encephalartos lebomboensis*, *Jubaeopsis caffra*, *Ficus natalensis* and *Albizia suluensis*, we assume that the specific names *capensis* (from the Cape), *lebomboensis* (from the Lebombo Mountains), *caffra* (from Kaffraria, i.e., the Eastern Cape), *natalensis* (from Natal, i.e., KwaZulu-Natal) and *suluensis* (from Zululand) all mean that these trees can be found growing in the areas mentioned. We need to go further with our assumptions in the following names: *Annona senegalensis* (from Senegal), *Ocotea kenyensis* (from Kenya), *Maerua angolensis* (from Angola), *Toddalia asiatica* (from Asia) and *Excoecaria madagascariensis* (from Madagascar).

We need here to assume that although Boon uses the phrases "from Senegal", "from Kenya", etc., he does not mean that these species were brought from these places and introduced to eastern South Africa. Were this so, they would be marked in his book as introduced species, of which there are indeed hundreds. We need to assume that, for example, *Annona senegalensis* grows naturally in both Senegal and eastern South Africa (and presumably in all the countries in between), but that it was in Senegal where the species was first

---

1. All examples here from Boon (2010).

discovered or first described. Occasionally, but only very occasionally, Boon gives us clues about this, as in *Acacia swazica* ("first collected in Swaziland" (2010: 186)) and *Commiphora zanzibarica* ("of Zanzibar, from where it was described" (2010: 216)).

In all these examples (*Cyathea capensis*, *Encephalartos lebomboensis*, *Annona senegalensis*, *Ocotea kenyensis*, etc.), the specific name is an adaptation of a toponym (place name). There are <u>no</u> such examples in the Zulu botonymical[2] system.

There are just a few botanical names that contain a locational reference to a habitat such as:

- *Croton <u>sylvaticus</u>* (pertains to forests);
- *Searsia <u>montana</u>* (a mountain habitat);
- *Ziziphus <u>rivularis</u>* (associated with small streams);
- *Gymnosporia <u>nemorosa</u>* (found in woods and groves);
- *Rhoicissus <u>napaeus</u>* (belonging to wooded dale or dell); and
- *Aloe <u>rupestris</u>* (rock-loving).

These names <u>do</u> have equivalents in the Zulu botonymical system.

(c) Commemorative names: *Dovyalis zeyheri* is the botanical name for the Apricot Kei-Apple. The specific name commemorates Carl Zeyher (1799–1858), a pioneer botanical collector in South Africa. In *Searsia lucida* (Glossy Currant) it is the <u>generic</u> name that commemorates Paul B. Sears (1891–1990), the head of Yale University's botany department. In *Searsia acocksii* (Pondo Climbing Currant) John Acocks (1911–1979), an "eminent botanist and ecologist", joins Professor Sears.

There are <u>no</u> such examples in the Zulu botonymical system.

Zulu plant name types with no equivalent in the botanical nomenclature

I am unable to find any examples at all in scientific botanical nomenclature of equivalents of the Zulu names **umhlalamagwababa** (where the ravens sit), the

---

2. The word 'botonymical' is a portmanteau word consisting of 'botany' + 'onymic' (pertaining to names). In creating this word for this book, I intend it to mean 'pertaining to a system of names for botanical entities'.

Coastal Goldenleaf (*Bridelia micrantha* (Pooley 1993: 218)); **isikhwelamfene** (where the baboon climbs), the Water Ironplum (*Drypetes arguta* (Pooley 1993: 212)) and the Pigeonwood (*Trema orientalis* (Pooley 1993: 66)); as well as **umdlampangele** (what the guinea fowl eats), the name for the Wild Pride-of-India (*Galpinia transvaalica* (Pooley 1993: 344)). We must conclude here that the botanists of the world who name plants do not see a need to refer to relationships between plants and animals, while in the Zulu botonymic system this is a regular and common feature.

Let us now look at the Zulu names in some detail.

Describing the plant itself

The 'maas' metaphor

The milky white fluid exuded by certain plants is metaphorically referred to as 'maas' (< *amasi* (sour milk)) in some plant names.

The Afrikaans and Zulu names for *Euphorbia bupleurifolia* share milk and sour milk metaphors. The Zulu name is **inkamasane** (what squeezes out a little maas) and the Afrikaans name is *melkbol* (milkball). Boon (2010: 504) gives the same Zulu name to the Forest Toad-Tree (*Tabernaemontana ventricosa*). Not only is the sap of this tree milky, but Boon says the sap is used to curdle milk.

English, Afrikaans and Zulu use the same idea in their names for *Sideroxylon inerme*. The Zulu name is **amasethole** (maas of the heifer), the English name Milkwood Tree and the Afrikaans name *witmelkhout*. Hutchings gives the same Zulu name to the Natal Bush Milkwood (*Vitellariopsis marginata*). Zukulu et al. (2012: 70) give *amasethole* as the Pondo name for the Pondo Milkberry (*Manilkara nicholsonii*), and say "*Amasethole* is a *hlonipha* name that may not be said aloud in the forest as it is believed that a leopard will appear before you and you will be in great danger" (2012: 231). This is a belief I have not come across at all in Zulu culture.

From Doke and Vilakazi (1958: 501) comes **ummfomasi** (< *mfoma* (ooze) + *amasi*), a name for an unidentified species of forest tree that oozes latex. The oozing of latex becomes the metaphorical weeping of latex in **inkalamasane** (what weeps maas), a name for *Euphorbia natalensis*.

Boon (2010: 502) gives **amasebele** (maas from the breast) as the Zulu name of the Forest Poison-Rope (*Strophanthus speciosus*), but to Doke and Vilakazi this is the name of the "**Euphorbia pugniformis* tree" (1958: 725).

Although Boon (2010: 230) does not mention any milky sap, Doke and Vilakazi (1958: 311) give the Zulu name of *Bridelia micrantha* as **umhlahlamakhwaba** (< *isihlahla* (tree) + *amakhwaba* (the sour milk of a strange homestead)).[3]

Boon simply gives the name **isibinda** (no underlying meaning) for the Wild Loquat (*Oxyanthus speciosus*), but for Doke and Vilakazi (1958: 78) the situation is a little more complicated. They distinguish between **isibinda esikhala amasi** (the *isibinda* tree that weeps sour milk) as referring to *Garcinia gerrardii* and **isibinda esingakhali masi** (the *isibinda* tree that does not weep sour milk) as referring to **Oxyanthus gerrardii* and the related *O. latifolius* and *O. natalensis*.

An interesting name is **amasikangcede**, for which Doke and Vilakazi give the literal meaning 'warbler's sour milk', and say "viscid exudation from the *ukhovothi* tree, used in making head-rings" (1958: 752). Note that this is not a plant name, but a name for the 'viscid exudation'. MacLean (1984: 592) identifies Doke and Vilakazi's 'warbler' as the Lazy Cisticola (*Cisticola aberrans*, Zulu *ungcede*). As for the identity of the '*ukhovothi* tree', Boon (2010: 68, 318) gives **umkhovothi** as a Zulu name for the Thorny Elm (*Chaetacme* [= *Chaetachme*] *aristata*) and *ukhovothi* as the Xhosa name for the Lemon Thorn (*Cassinopsis ilicifolia*), but for neither tree does he mention any 'viscid exudations'.

From Doke and Vilakazi we also get the name **amasamunyu** (sour maas), which they say refers to a "species of forest tree, prob[ably] *Cassipourea verticillatum*" (1958: 722). As we might expect, 'Cassipourea verticillatum' (correctly **Cassipourea verticillata*') is not in Boon, but it is probably the tree he refers to as *Cassipourea gummiflua* var. *verticillata*. In his description (2010: 396) there is no mention of maas-like white, milky sap, but he does point out that 'gummiflua' means 'dripping with gum', so perhaps this is the origin of the metaphor of sour milk that is even more sour than usual. The names of this tree are discussed in greater detail under the sub-heading 'Smell or taste' in this chapter.

And to round off this section on maas-metaphors in Zulu plant names, here is an interesting note from Gerstner:

---

3. Zulu traditions, and the Zulu language, make a distinction between *amasi*, the sour milk of your own homestead, which you may drink; and *amakhwaba*, the sour milk of any other homestead, which you may not drink.

umHlambamasi . . . 'The sour-milk cleaner', *Rauvolfia natalensis*
. . . the Quinine tree, with big panicles of little white flowers and
whorled shiny leaves, which, when cut, exudes latex. The bark is used
as emetics in fevers, when plenty of sour milk will no longer stay in
the stomach. Hence the name (1938: 136).

Behaviour and growth patterns of plants

Let us start with the Zulu name of the Triffid Weed (*Chromolaena odorata*),
described by Boon as "one of the most invasive plants in KwaZulu-Natal"
(2010: 582): **usandanezwe** (< *anda* (increase) + *na* (with) + *izwe* (country,
land; i.e., the plant which is increasing over the whole country)).

The seeds of some plants are dispersed by sticking to the hide, skin or
clothes of passing animals and humans. This is referred to in the following
names.

The name **umhlabangubo** (what stabs the clothing) refers to the Common
Blackjack (*Bidens pilosa*), while **isinama** (< *namathela* (stick to)) is a name
for *Desmodium incanum*. Pooley says of this plant that the fruits are covered
in sticky hairs and so dispersed by people and animals (1998: 394). This habit
of sticking closely to passers-by is delightfully captured in the English name
for this plant – Sweethearts!

Both **isinama** and **isibambangubo** (what catches the clothing) are names
for the Burrweed, *Achyranthes aspera*. The Afrikaans name *haak-en-steek-
klitsbossie* also refers to this characteristic of the plant, described by Pooley as
"[fruits have] starry spines which attach to people and animals" (1998: 528).

The verb *namathela* (stick to) is used on its own with a noun prefix in the
name **inamathela** (what sticks to things), a reference to the Sticky-Leaved
Monopsis (*Monopsis stellarioides*).

The Zulu, English and Afrikaans names for *Myrothamnus flabellifolia* all
have the same underlying meaning: the Zulu name is **uvukakwabafile** (what
rises among the dead), the English name Resurrection Plant and the Afrikaans
name *opstandingsplant*. Pooley says "In dry weather the whole plant withers
and appears to die. [It] revives within hours after rain" (1998: 576). Hutchings
says of the genus as a whole "plants have leaves which fold up when desiccated
but rapidly revive and stand erect after rain" (1996: 116).

The White Stonecrop (*Crassula vaginata*) has the Zulu name
umakhulefingqane (it grows contracting itself into a mass; or, it grows
drawing itself together). The White Milkwood, mentioned in the section on

maas-metaphors, has the similar name **umakhwelafingqane** (what climbs crookedly), a reference to the crooked growth of the main stem.

In Chapter 5 we came across the name **imbuna** (what droops), a reference to the Sensitive Plant (*Mimosa pigra*) of which Pooley says "leaves collapse on touching" (1998: 388).

Sarcophyte sanguinea is a plant that just seems to pop up out of the ground, according to its Zulu name **umavumbuka** (the one that pops up). From popping up out of the ground we go to creeping along the ground and look at the name **ijingitheka**, which Doke and Vilakazi give as a name for a "species of herb, *Vernonia hirsuta*" and say it is derived from the verb *jingitheka*, meaning "creep along ground, as plant" (1958: 361).[4] Pooley (1998: 498) gives 'ijungitheka' [*sic*] as one of the four Zulu names for this plant, but says nothing about it creeping along the ground. From creeping along the ground, to creeping over the dunes: Doke and Vilakazi give the name **unqwambane** for a "species of Verbena, *Vitex harveyana*, which commonly climbs over sand dunes" (1958: 597) and say the name is derived from the verb *nqwamba* (be draped over the shoulders). There is, however, no mention of this dune-climbing habit in Boon's description of *Vitex harveyana*, which he says is a "small tree . . . in thicket and woodland" (2010: 520). Nor does Boon recognise the name **unqwambane**, giving the Zulu name **umluthu**.

Sprawling, crawling, climbing or creeping – these seem to be plant characteristics that manifest easily in plant names. Doke and Vilakazi give the name **umthambalazi** (< *thámbalaza* (sprawling)) and say it is a "species of climbing plant, *Asparagus sprengeri*" (1958: 782). Pooley (1998) refers to twenty different species of the genus *Asparagus*, but unfortunately 'Asparagus sprengeri' is not among them. Hutchings (1996), usually so helpful about giving earlier synonyms, is unable to help either, so the sprawling or climbing nature of **umthambalazi** must be left unclear.

Also from the pages of Doke and Vilakazi (1958: 784) comes **intandela** (< *thandela* (wind around, twine)) given for any creeping or climbing plant.

---

4. Beaumont (personal communication) points out that this is not really a creeping plant, so one wonders about Doke and Vilakazi's derivation from the verb *jingitheka*.

We end this sub-section with a query: what kind of plant behaviour is referred to in the name **uhlinzamfuka**, which Doke and Vilakazi interpret as 'what cuts open the thicket?' They say, rather imprecisely, that this refers to the "common meadow fern" (1958: 329). Neither Pooley (1998) nor Hutchings (1996) recognises the Zulu name.

Thorns

That there are a number of plant names which specifically relate to thorns must be attributed to the fact that thorns have a way of impinging strongly on human consciousness. Whether one is collecting firewood, picking fruit or simply walking through the veld, thorns make their presence known. The Hairy Turkey-Berry (**Canthium mundianum*) has the most semantically transparent of such names, with the Zulu name **umevane** (< *ameva* (thorns) + the plant-name forming suffix -*ane*). Boon describes this tree as having "paired, straight, slender spines at right angles" (2010: 540).

A number of these 'thorn-names' are based on the verb *hlaba* (stab, prod) and related verbs. Two specific bird species are the named victims of this stabbing and prodding in **umhlabandlazi** (what stabs the mousebird), the Tree Aloe (*Aloe barberae*), and **umhlabahlungulu** (what stabs the crow), the Tassel Berry (*Antidesma venosum*). The pairing of thorns is emphasised in **uhlabazihlangane** (they stab being together), the name for the Prickly Red-Berry (*Erythrococca berberidea*) of which Pooley says: "Spines short, straight, yellow, in pairs" (1993: 224). From the ideophone *dlóvu* (of stabbing) comes the name **umdlovune**, the Fever Tree (*Acacia xanthophloea*) and the verb *hlokoloza* (prod, poke, jab) is the base of **isihlokolozane**, the Lemon Thorn (*Cassinopsis ilicifolia*).

The names above are all of thorny trees. Some smaller plants that prick and stab just as heartily as their bigger cousins are referred to in the following names, most of them based on the verb *hlaba*.

Aloe gerstneri has the Zulu name **isihlabana** (stabbing). Pooley states "young leaves spiny on both surfaces" (1998: 32). The use of the suffix -*ana* on the verb *hlaba* here, and in the next example, indicates a continuous action. The verb *hlaba* is repeated in the name **umahlabahlabane** (a continuous little stabbing), a name given to *Kyphocarpa angustifolia* of which Pooley says: "leaves . . . sharp tipped . . . bracts spine tipped" (1998: 134).

The Zulu name **isihlabamakhondlwane** (what stabs like spears) refers to *Dicoma anomala*. Doke and Vilakazi say that this word refers to "several spp. of *Dicoma*, a silver thistle with very spiky bracts" (1958: 309).

Doke and Vilakazi give **umhlabankonkoni** (what stabs the wildebeest) as referring to a "species of tree, prob. Rhus Legatii [*sic*]" (1958: 309), but Boon gives **umhlabankonkani** (note different spelling) as a Zulu name for the Forest False-Spikethorn (*Putterlickia verrucosa*) with spines up to 70 mm long (2010: 314). For the same tree, Boon also gives the name **uhlinzanyoka** (what skins a snake), which Ndela Ntshangase tells me refers to thorns.[5]

Aloe suprafoliata (spiny leaves) goes by the Zulu name **umhlabandlazi** (what stabs the mousebird). Why it should be the mousebird specifically that is stabbed by the thorns of *Aloe suprafoliata* is not clear, any more than it should be the grasshopper in **uhlomantethe** (what impales the grasshopper) a name for *Indigastrum fastigiatum*. Doke and Vilakazi state: "species of Indigofera shrub . . . on which grasshoppers are often found impaled" (1958: 333). Ndela Ntshangase adds that this is usually done by the bird *iqola* (*Lanius collaris*, the Fiscal Shrike or Jacky Hangman).[6]

Metaphors based on the similarity between thorns and horns are the basis of **umpondonde** (long horns), a name for the Tree Aloe (*Aloe barberae*), and **umngampondo** (*umunga* (acacia) + *izimpondo* (horns)), a reference to *Acacia grandicornuta*. The English name for this species is Horned Thorn and the Afrikaans name is *horingdoring*, so all four names for this plant use the same metaphor.

The well-known Afrikaans name *wag-'n-bietjie* (wait a bit) suggests that the tree is attempting to stop or delay the passer-by and this same notion is reflected in Zulu 'thorn names' in the following:

- **ubophanyamazane** (what arrests the buck): a name for both the Water Ironplum (*Drypetes arguta*) and the Magic Guarri (*Euclea divinorum*);
- **ubophe** (arrest, stop): a reference to the Flame Thorn (*Acacia ataxacantha*) with the alternative name **umthathawe** (< -*m*- (him) + *thatha* (take) + *awe* (and he falls; i.e., take him so he falls)), a reference to the thorns catching at the ankles of a passer-by);
- **umbambampala** (catch the impala), which refers to three acacias: the Black Monkey Thorn (*Acacia burkei*), the Knob Thorn (*A. nigrescens*) and the Bastard Umbrella Thorn (*A. luederitzii*); and

---

5. Personal communication.
6. Personal communication. Ntshangase says the bird usually leaves the locust to rot on its thorn, attracting maggots (*impethu*) that the bird can eat at its leisure.

- **umbambanhlangu** (what catches the reedbuck): the Zulu name of the Red-Heart Tree (*Hymenocardia ulmoides*). As regards this last tree, Pooley has no mention of thorns, but states "leaves heavily browsed" (1993: 214), so possibly it is the nutritious appeal of this tree that delays the reedbuck rather than thorns.

The notion of heavy growth seems more likely in the Zulu names **ithiyampondo** (what catches the horns) and **ithiyela**. Both are based on the verb *thiya* (trap, hinder, obstruct). Doke and Vilakazi give the first as a name for an unidentified species of forest tree and the second as a "species of composite shrub, *Cassinia phylicifolia" (1958: 797). Pooley, Boon and Hutchings do not recognise the genus 'Cassinia'.

The verb *sondela* (approach), together with its transitive derivative *sondeza* (cause to approach), is the basis of **usondeza**, the Wild Caper Bush (*Capparis sepiaria*), with its "strong, paired, hooked thorns" (Pooley 1993: 104) and **usondela**, **usondeza**, the Cat Thorn (*Scutia myrtina*), which has "hooked thorns, usually in pairs, sharp" (Pooley 1993: 300). This tree also has a name that combines the verb *sondela* with the verb *anga* (to kiss) to produce what is certainly my favourite tree name in the Zulu lexicon: **usondelangange** (come closer so that I can kiss you).[7] Pooley (1993: 90) also records **usondeza** as the name of the False Bougainvillea (*Pisonia aculeata*) with its spiny stems.

Finally, while *inkalimeva* (< *bukhali* (sharp) + *ameva* (thorns)) is not a plant name, it is interesting to note the thorn metaphor in this name for a "fabulous cat in witchcraft beliefs" (Doke and Vilakazi 1958: 377). Clearly witchcraft, even in the form of a cat, can stab badly.

Tough, hard wood

This category overlaps with Chapter 9 as tree names referring to the hardness or toughness of wood are directly related to the practical use humans make of the wood, whether for firewood or for the purposes of building or manufacture. Many of the names, indeed, refer to such human activity by describing the wood as being "so tough as to defeat an axe" as in **umhlulambazo** (what overcomes the axe), the name for the Natal Box (*Buxus natalensis*). Similar names are **umzilazembe** (what shuns the axe), which refers to the Sickle Bush

---

7. *sondela* + *ngi* (I) + subjunctive form of *anga* (kiss): 'approach that I might kiss'.

(*Dichrostachys cinerea*), **umdlulamazembe** (surpasses the axes), the Common Pheasant-Berry (*Margaritaria discoidea*) and **umhlabambazo** (what stabs the axe), the Dune Bride's Bush (*Pavetta revoluta*).

Simple references to hard, tough wood are **umazwenda** (< *uzwenda* (tough object)), the name of the Red Hook-Berry (*Artabotrys monteiroae*); **umuthinzima** (hard tree), the False Soap-Berry (*Pancovia golungensis*); and **umanzimane** (< *nzima* (tough)), the Natal Worm Bush (*Cadaba natalensis*).[8] For Doke and Vilakazi, **umanzimane** is the name of "a species of shrub and tree, **Royena lucida*" (1958: 485).

The metaphor of bone is found in **ithambo** (bone), a reference to the tough wood of the Galla Plum (*Haplocoelum gallaense*) and also a name for the Forest Elder (*Nuxia floribunda*) with its "pale yellowish white, hard [wood]" (Pooley 1993: 422). The same metaphor is linked to 'hyena' in **ithambolempisi** (hyena's bone), the Zulu name of the Common Canary-Berry (*Suregada africana*). It may be that this last name is linked to the known fact that hyenas, with their immensely strong jaws, are able to crack the hardest bone. Similar to the bone metaphor is the ivory metaphor. The Leadwood (*Combretum imberbe*) has as one of its four Zulu names the name **impondondlovu** (elephant's tusk). The wood of this tree is exceptionally hard and heavy (cf. English common name).

Another metaphor, this time of iron, is the basis of **umsimbithi** (< (*in*)*simbi* (iron) + (*umu*)*thi* (tree)), the name of *Millettia grandis*. We have discussed this name in detail in Chapters 1 and 3. The diminutive form of **umsimbithi** is found in **umsimbishana** (**Millettia sutherlandia*), which curiously is named in English the <u>Giant</u> Umzimbeet (Pooley 1993: 166). There is a puzzle here: why should Zulu botanists regard *M. sutherlandia* as a small version of *M. grandis*, while English-speaking botanists regard it as a giant version? Pooley gives the answer: *Millettia sutherlandii* (the Giant Umzimbeet) is considerably bigger than *M. grandis*, but has much smaller flowers.[9] So both vernacular names are correct, with the English vernacular referring to the tree as a whole and the Zulu referring to the flowers.

We saw earlier in Chapter 3 that Van Warmelo, on the basis of the Venda name *Mutomboti* for the hardwood tree **Canthium mundianum*, regarded 'thombo' in the Zulu tree name **umthombothi** (*Spirostachys africana*) as being

---

8. Not found in Boon (2010).
9. Personal communication.

an obsolete Zulu word for 'stone', giving us 'the tree with wood hard as stone'. If this is so, and 'thombo' is <u>not</u> derived from *umthombo* (fountain, spring) then iron, stone, ivory and bone are all metaphors for hard, heavy wood in the Zulu dendronomasticon.[10]

Smell or taste of plant

Many plants are strongly scented, either with a pleasant scent, or an unpleasant, offensive smell. First, plants with unpleasant smells.

Our first example is **umafuthasimba** (< *amafutha* (fat) + *isimba* (dung or excrement)), a Zulu name for the herbaceous climber *Kedrostis foetidissima*.[11] Pooley (1998:78) says it has an unpleasant smell when broken. Also using excremental imagery is the name **umkaka** (< *amakaka* (excrement of babies)) for the foul-smelling African Cucumber (*Momordica balsamina*).

The name **umbola** (< *bola* (to rot)) is used for both the unpleasant smelling Forest Pineapple-Flower (*Eucomis bicolor*) and the Natal Candelabra Flower (*Brunsvigia natalensis*) of which Pooley says "[the] Zulu name refers to the rotten smell around the bulb" (1998: 348).

Not as unpleasant, but still strong smelling, is the Tall Khaki Weed (*Tagetes minuta*) with its "pungent smelling leaves" (Pooley 1998: 318). Its name **umnukani** (smells of what?) is also used for the Black Stinkwood (*Ocotea bullata*) and the Sandworm Plant (**Chenopodium ambrosioides*) where the question posed by the Zulu name (smelling of what?) is answered in the Afrikaans name *hondepisbos* (dog urine bush) with Pooley saying mildly "the plant has a characteristic scent" (1998: 526). The name can also be used for pleasant smells: **umnukani** is a name for the Water Mint (*Mentha aquatica*), which Pooley tells us is a sweet-smelling fragrant herb (1998: 424).

The Zulu name of the Sweet-Reed (**Andropogon sorghum*), **imfe**, is the basis of **imfenyane** (little sweet-reed), the *ruikbossie* (*Senecio rhyncholaenus*) with its "characteristic sweet or spicy smell" (Pooley 1998: 328). Note that the Afrikaans name means 'little smelly bush'.

---

10. I have coined this word from Greek roots, with the meaning 'list of names referring to trees', in line with terms like *anthroponomasticon* (list of personal names) and *toponomasticon* (list of place names), both of which are regularly used by onomastic scholars.
11. The Latin name *foetidissima* means 'extremely smelly'.

The epiphyte *Cyrtorchis arcuata*, which is "sweetly-scented at night" (Pooley 1998: 130),[12] has the Zulu name **imfeyenkawu** (sweet-reed of the monkey). Ndela Ntshangase says monkeys are well known for their liking for this plant.[13]

Other plants with strong scents reflected in the name include:

- **umnandi** (pleasant): the lemon or clove scented Wild Basil (*Ocimum gratissimum* (very pleasing);
- **utshwalabezinyoni** (beer of birds): the Wild Dagga Plant (*Leonotis leonurus*); and
- **ishaladi lezinyoka** (garlic/shallots of the snakes): the Wild Garlic (*Tulbaghia acutiloba*) of which Pooley says "plant parts smell of garlic. Used . . . as a snake repellent" (1998: 510). This name is very similar in its underlying meaning to **unqandanyoka** (what keeps away a snake), which Doke and Vilakazi say refers to "a species of Senna weed, *Cassia occidentalis*, with a very repugnant smell" (1958: 590). Hutchings (1996: 130) gives *Cassia occidentalis* as an earlier synonym for *Senna occidentalis* (or Stinkweed). She also offers the Zulu names **isinyembane** and **umwanda-nyoka** for this plant. The snake (*inyoka*) is easily recognisable in the latter name, but Doke and Vilakazi do not recognise *umwanda* or *wanda* and give *isinyembane* as being the "language [and] characteristics of Natives of Inhambane" (1958: 624), which seems a little far removed from the world of plants.

The 'names of odour' described above were mainly of smaller plants. Let us now turn our noses to trees. Tree names that refer to the smell of certain parts of the tree concentrate on unpleasant smells. The smell may be specifically identified, often by metaphor, as is the case with **isidumbu** (corpse), the Tonga-Kerrie Tree (*Cladostemon kirkii*) with its "pervasive, peculiar, unpleasant odour when picked" (Pooley 1993: 106);[14] or the name might be vague, as is the case with **isidadada** (evil-smelling thing), the name of the Skunk Bush (*Premna mooiensis*), which Pooley describes as having a pungent, unpleasant

---

12. Beaumont points out (personal communication) that many plants have sweetly smelling white flowers that are most fragrant at night, attracting night-flying pollinators like moths. White flowers are also the most visible at night.
13. Personal communication.
14. Ndela Ntshangase says (personal communication) that this tree is also known as **umusi-iyamuka** (the smell is moving).

scent (1993: 440). Ndela Ntshangase says of this last-named that it also has the name **umqathathongo** (what causes friction among spirits), suggesting potential for witchcraft.[15]

The reference to the smell of "an emission of intestinal gas from the anus"[16] is clear in two plant names given by Doke and Vilakazi (1958: 771), both based on the verb *suza* (fart). The name **umsuzane** is given for an unidentified species of strong-smelling shrub, while **umsuzwane** is more specific as it is a "species of shrub, **Lippia asperifolia*, having a very disagreeable smell, used for smearing the body as a protection against crocodiles and dogs". Hutchings (1996: 263), who gives *Lippia asperifolia* as an earlier synonym for *Lippia javanica*, confirms the practice. Pooley mentions the use of this plant for ritual cleansing after contact with a corpse, and also as protection against dogs, crocodiles and lightning, but raises doubt about the use of the farting metaphor in the two Zulu names in her comment "[used as] a fragrant cupboard freshener" (1998: 180). The Zulu names also seem somewhat at odds with the English common name Lemon Bush.

Evil smell and habitat are combined in **umnyamanzi** (< **umunya** (sp. of evil-smelling plants) < *nya* (excrete) + *amanzi* (water)), which refers to the Large-Leaved Onionwood (*Cassipourea gummiflua*) that also has the name **isinuka** (smelly thing). It is, however, possible that the name **umnyamanzi** refers to a tree that produces water, for Boon (2010: 396) also gives the name **umjuluka** (sweaty thing) for this tree. Boon does not mention anything about the tree producing water, but there is a clue in the specific name *gummiflua*, which means 'dripping with gum'. All in all, this sweaty, smelly, water-excreting, gum-dripping tree sounds like rather an unpleasant tree to sit under.

The striped field mouse (*Rhabdomys pumilio*) clearly has an odour that relates strongly to the smell of certain trees and two tree names are based on the verb *nuka* (smell) and the noun *imbiba*, the Zulu name for mouse.

The first Zulu name is **umnukambiba**, which means 'smell of striped field mouse', but the Latin, English and Afrikaans names for this tree have their own metaphors: *Clausena anisata*, Horsewood and *perdepis*. Pooley comments: "Leaves smell very strongly of aniseed, but the Afrikaans name 'perdepis' is suitably descriptive" (1993: 192). Boon (2010: 204) also gives the name **umsanga**, of which Doke and Vilakazi say "Species of shrub,

---

15. Personal communication.
16. This is how *Collins Dictionary of the English Language* (1986: 551) defines 'fart'.

Clausena inaequalis [an earlier name for *C. anisata*]" (1958: 722). Under the spelling **umsanka**, however, Doke and Vilakazi give the meanings "species of strong-smelling shrub, *Clausena inaequalis*, used to expel vermin" and "strong, disagreeable smell of the body when dirty and perspiring" (723). So from a sweaty, smelly, water-excreting, gum-dripping tree, we have gone to a tree that simultaneously smells of a striped field mouse, horse urine, aniseed, and a dirty and perspiring human body. I wouldn't want to sit under this tree either. Ntshangase adds, by the way, that parts of this tree should not be brought back to the homestead as they can be used to make *isichitho*, associated with witchcraft.[17]

The second name based on the striped field mouse, *imbiba*, using the verb *nuka* in its applied form *nukela* (smelling of), is **umnukelambiba**, used for both the Cats Whiskers (or Verbena Tree, **Clerodendrum glabrum*), which has a "pungent smell when crushed" (Boon 2010: 514) and the Sickle Bush (*Dichrostachys cinerea*). In the absence of further evidence, we must assume that these two trees only smell of striped field mice, in other words, not of horse urine, aniseed, sweaty bodies or any other smelly entity.

Coming now to taste rather than smell, we find **itswayilentaba** (salt of the mountain), the Salt of the Shepherds (*Epilobium hirsutum*). Pooley notes that "herdsmen lick the plant for the salty taste" (1998: 410).

Unpleasant tastes found in tree names are occasionally based on the adjectives -*muncu* and -*munyu*, both of which mean 'sour', 'acidic' and 'bitter'. In **isimuncwane**, the Gland-Leaf Tree or Large-Leaved Bride's Bush (*Pavetta edentula*), -*muncu* is used with the suffix -*ane*. The sour taste of certain parts of this tree is also expressed in the ominous name **umafayindlala** (die of hunger), no doubt because the bitter taste makes it unsuitable for eating. Surprisingly enough, one of the Swazi names for this tree given by Boon (2010: 560) is *sawoti* (salt).[18]

In **umunyumunyu**, the name for the Climbing Raisin (*Grewia caffra*), -*munyu* is reduplicated with augmentative effect: 'sour-sour'. *Munyu* is used with the plant-name-forming suffix -*ane* in **umunyane**, which Boon (2010: 532) gives as a name for the Cape Honey-Suckle (**Tecomaria capensis*). Doke and Vilakazi take this name a lot further with the following: **u(lu)munyane**: "sp. of plant with sour [but] edible leaves, cf. *isimunyane*" (1958: 517); **isamunyane**:

---

17. A charm used to cause discord in the family, or estrangement, separation and divorce.
18. The other, not unexpectedly, is *simunyane*.

"veld plant with sour edible leaves, *Pavetta geniculata*, eaten by travelers and herd-boys as a refreshing tonic" (1958: 8); and (based on -*muncu*) **isimuncwane**: "certain plants with acid leaves, *Oxalis semiloba*, *O. smithiana*, *O. corniculata*, cf. *isimuncwane*, *isamunyane*" (1958: 516).

Earlier in this chapter, when talking about metaphors based on sour milk, we mentioned the name **amasamunyu** (literally, sour milk which is sour), another Zulu name for the Large-Leaved Onionwood (**Cassipourea gummiflua* var. *verticillata*).

Equally unpleasant to the taste is the plant **ishaqa** (also **ushaqa**). Doke and Vilakazi say this is a "species of astringent herb, *Berkheya* species, also **Pelargonium aconitophyllum*" (1958: 733) and that the name is derived from the verb *shaqa* (be astringent, acrid, acid, pungent). Without knowing which of Pooley's (1998) eighteen *Berkheya* species is being referred to, it is difficult to comment on the degree of astringency of pungency of the **ishaqa** plant. Hutchings makes no mention of pungent or acrid taste in her descriptions of the medicinal use of several *Berkheya* species, apart from saying that the roots of an unidentified *Berkheya* species are pounded with leaves in cold water and used as a "cold stringent" for sores (1996: 333). Hutchings (1996: 149) gives **Pelargonium aconitophyllum* as an earlier synonym for *Pelargonium luridum* (Wild Geranium) and both **ishaqa** and **ishaqwa** as Zulu names for the plant. There is no indication, however, of pungent or acrid qualities mentioned in her description of the medicinal uses of this plant. We do note, though, that its dried powdered roots are mixed with hippopotamus or python fat and rubbed on the face by young men as a courting charm. We shall meet this plant again in Chapter 8 when we shall discuss the alternative Zulu names **inyonkulu** (big bird) and **isandlasonwabu** (hand of the chameleon).

From bitter to pleasant tastes. Two trees have names based on the adjective stem -*mnandi* (pleasant tasting): **umnandana** (fairly pleasant tasting), a name for the Common Pheasant-Berry (*Margaritaria discoidea*) and **umnandi** (tasty), Natal Plane (*Ochna natalitia*). From Doke and Vilakazi (but not in Pooley) is **umaphungwana** from the diminutive plural of *iphunga* (scent, fragrance), used for "[any] tree of the Croton species" (1958: 485).

One might expect a plant named **umoyawentombi** (maiden's breath) to be sweet to the nose, so it is surprising to see that Doke and Vilakazi give this as the name of a "species of carrion flower" (1958: 508). In Pooley's entry on *Stapelia gigantea* (Giant Carrion Flower) there is no mention of smell, but she does say "used in sorcery as a poison, reportedly capable of causing death" (1998: 302) and somehow this seems an unsuitable feature of 'maiden's breath'. However, in her description of Leendertz's Carrion Flower (*Stapelia leendertziae*) we find the remark "very disagreeable foetid smell" (1998: 584). The plant *Orbea woodii* (previously *Stapelia woodii*) is described as "smells like human faeces" (1998: 586). These descriptions help explain Doke and Vilakazi's reference to a 'carrion flower', but **umoyawentombi** remains, alas, unexplained. Smith has:

> **carrion flower.** Various species of Stapelelieae, see AASBLOM. The vernacular name is derived from the putrescent carrion-like odour emitted by the flowers of many of the species, and from their suggestive colouring (1966: 189).

Interestingly, although Smith explains the word 'aas' as "the original Dutch means bait,[19] with the distinct meaning of carrion-like or of having an unpleasant, even foetid odour" (1966: 56), he gives the vernacular name *aasbossie*, used for the plant *Coleonema album*, and says "the <u>aromatic</u> [my emphasis] leaves are used by the fisherfolk of the L'Agulhas coast to cleanse their hands of the vile smell of red bait."

Appearance of plant

Colouration

Some of the more basic plant names that refer to the appearance of the plant are those relating to colour as in **uhlamvuhloshane** (whitish leaf),[20] a name for the Pink-and-White Gerbera (*Gerbera ambigua*).

The Zulu colour term *-luhlaza* refers to both green and blue and it is used for green and blue independently in the following two names: **ikhakhakha eliluhlaza** (green thistle), a name for Dainty <u>Green</u> Bells (*Dipcadi viride*), with its bright emerald-green flowers, and **ikhambi eliluhlaza** (blue herb), a name for <u>Blue</u> Stars (*Aristea ecklonii*), which has bright blue flowers.

Red flowers give rise to the following three names:

- **untanganazibomvu** (young red folks): the bulbous herb *Ledebouria ovatifolia*;

---

19. Beaumont (personal communication) finds the word 'bait' very appropriate. She writes: "These flowers attract fly pollinators. Flies lay their eggs on the flowers. They are duped into thinking the flower is carrion. The fly eggs and larvae are doomed to die, as there is no animal flesh on which they can feed."
20. The hairs on the leaves give them a whitish appearance.

- **uvemvane olubomvu** (red butterfly): the Little Russet Pea (*Argyrolobium tuberosum*) where the little red flowers are in the form of matching 'wings'; and
- **ihlule** (blood clot), the parasitic *Sarcophyte sanguinea*. The little red 'blobs' of flowers look just like clots of blood and the Latin name *sanguinea* means 'blood red'.

The blood metaphor is common for red colours and found in the common Zulu name for a red ox, *uJamludi*, a name derived from the Dutch *Jan Bloed* (Jack Blood). While on the subject of blood, we can note the name **ungazi** (also **ungazinde**, **ungazini**, all derived from *i*[*n*]*gazi* (blood)). Doke and Vilakazi give these three names for "a species of forest tree whose bark exudes a red sap, said to cause pupurea and fatal haemorrhages when used by witches" (1958: 558). These names cannot be found in Boon, Hutchings or Moll. It is possible, though, that this is the Kiaat (*Pterocarpus angolensis*) for which Boon gives the Zulu names **umvangazi** and **umbilo** saying of the tree that "Sap [is] blood red, [and is] used for magical & medicinal properties and dye" (2010: 166).

The Zulu name for the Little Russet Pea, as we saw above, combines -*bomvu* (red) with the metaphor of a butterfly. The same metaphor is seen in the name **uvemvane samatshe** (butterfly of the rocks) for the Necklace Vine (or String of Hearts, *Ceropegia woodii*). The Zulu name refers to the 'double-wing' shape of the young leaves. Another insect metaphor is found in **isinambuzane** (insect), a name for the Lesser <u>Moth</u>-Fruit Creeper (*Sphedamnocarpus pruriens*) with fruit that appears to have little wings (see illustration on page v).[21]

Tree names that refer to colouration, usually of the bark or the leaves, run from light to dark with shades of yellow and red in between. Light-coloured stem and leaves are referred to in **umhloshazane** (< -*mhlophe* (white)), the name of the Forest Fever-Berry (*Croton sylvaticus*). According to Doke and Vilakazi, this name refers to "one of a number of plants with whitish leaves or stem" (1958: 336).

The diminutive form of the relative stem -*mpofu* (tawny yellow) is seen in **umpofana** (yellowish), a name for the False Soap-Berry (*Pancovia golungensis*), while a reddish colour is referred to in the names **umbovu** (red

---

21. Beaumont (personal communication) states that these winged fruits are called 'samaras' and their shape suggests a moth at rest.

one) or **umbhovane** (reddish one),[22] both used for the Natal Plane (*Ochna natalitia*), a tree that would also appear to be known as the Mickey Mouse Bush, perhaps because the fruits look like the ears of Mickey Mouse.

Dark foliage is referred to in **umnyamathi** (< *umnyama* (dark) + (*umu*) *thi* (tree)), the name of the Cape Ash (*Ekebergia capensis*) with its very dark green foliage. This tree has an alternative name in **umathunzini-wezintaba** (in the shadows of the mountains). The idea of dark foliage being in permanent shadow is also seen in **umathunzini** (in the shadows), a name referring to both the Forest Mahogany (*Trichilia dregeana*) and the Natal Mahogany (*T. emetica*), both with very dark green foliage. Ndela Ntshangase suggests[23] that all these names I have interpreted as relating to dark foliage in fact refer to a 'dark habitat' as all these trees grow in deep forest. His suggestion would certainly be likely for **umlalamnyama** (what stays in darkness), which Doke and Vilakazi (1958: 446) give as a name for a tree of the *Euclea* genus.

The names for two species of *Cassinopsis* combine the adjective -(*lu*)*hlaza* (green) with the words for 'snake' and 'mamba' in order to give a picture of a slim, green creeper that climbs into the forest canopy: **imamba eluhlaza** (green mamba), the Lemon-Thorn (*Cassinopsis ilicifolia*) and **ihlazane** (greenish), also **inyoka-eluhlaza** (green snake), the False Lemon-Thorn (*Cassinopsis tinifolia*). This last named also has the name **inyandezulu** (lit., bundle of sky). This name is also used for a species of green snake, according to Doke and Vilakazi (1958: 620). Boon gives both **imamba-eluhlaza** and **inyandezulu** for this tree, now called *Cassinopsis ilicifolia*, and says "Zulu name *inyoka-eluhlaza* [which in fact he does not list as a Zulu name for the species!] means green snake, refers to colour of the stems" (2010: 318).

The light-coloured bark that makes the tree stand out from others at a distance is featured in **umkhanyakude** (what shines from afar), a name for the Fever Tree (*Acacia xanthophloea*) (see Plate 11). This Zulu name is shared with the Septee Tree (*Cordia caffra*). Pooley says of this tree that it is "conspicuous in [the] forest" (1993: 438) probably because of its bark, a light-coloured creamy-pink.

---

22. The Zulu word -*bomvu*, also found as -*bovu*, does not refer to a scarlet, carmine or crimson red, but more of a russet.
23. Personal communication.

Other descriptions

Straightforward, non-metaphorical description of leaf and flower is found in the two names **umagobongwane** (little hollow things), the Christmas Bells (*Sandersonia aurantiaca*), with flowers that look like little hollow bells; and **uqabikhulu** (big leaf), the Wild Basil (*Ocimum gratissimum*).

On the other hand, metaphors are found in:

- **uthangazane omncane** (smallish pumpkin): a name for the Wild Cucumber (**Coccinia palmata*). Both the Zulu name and the Afrikaans names *bospampoentjie* and *wildepampoentjie* perceive the resemblance of the fruit of this plant to a small pumpkin;
- **insangwana** (little cannabis plant): the Tall Khaki Weed or Mexican Marigold (*Tagetes minuta*), which has leaves with toothed outlines very similar to those of the cannabis plant (Z. **insangu**); and
- **ikhwanyana** (little sedge): the name of Blue Stars (*Aristea ecklonii*).[24] Another name for this plant is **umafosi** (< plur. *ifosi* (tip of whip)), a reference to the black-marked tip of the long lanceolate leaves.

The hairy quality of fruit is the basis of the name **isikhukhuboya**, which Doke and Vilakazi explain as derived from **isikhukhu** (species of rough-leaved grass growing in damp places) + *uboya* (fur, hair on animal). They give the name **isikhukhuboya** as an alternative for the *isikhukhu* grass and also for a "species of wild fig-tree, with small, hairy fruit" (1958: 410). Boon gives this as the name for the Sycomore Fig[25] (*Ficus sycomorus*) with its "large, round . . . finely hairy" fruits (2010: 80). Curiously enough, it is the Swamp Fig (*Ficus trichopoda*) with the specific name (*trichopoda*) which means 'hairy feet'. The hairy feet of this tree are not reflected, though, in the Zulu name **umvubu** (< *imvubu* (hippopotamus)). Boon says of this tree "First described by Rev J. Gerstner, missionary and botanist from Zululand, who named it after the hippo whose distribution was similar to the tree, the animal & tree have

---

24. Beaumont (personal communication) suggests that the brown bracts surrounding the flowers may resemble the inflorescences of sedge leaves.
25. I am not sure why Boon uses this spelling. In his description of the tree he says "Sycomore of Bible. True sycamore is *Acer pseudoplatanus*". Note also the botanical name, with the generic name being the Latin for 'fig', and the specific name derived from the Greek *sykon* (fig) with the Latin *morus* (mulberry). Taken together, these give the tree a botanical identity of 'fig-fig-mulberry'.

the same Zulu name" (2010: 80). This information comes, via Pooley (1993) from Palmer and Pitman (1972 vol. 1: 462) who identify this tree as "*Ficus hippopotami* Gerstn." and give the vernacular names Hippo Fig, wild rubber tree and **umVubu**. Palmer and Pitman say:

> The hippo fig is a common tree of the coast and swamp forests of Zululand. Its distribution in Zululand is very roughly that of the hippopotamus and the Zulus use the same word (umVubu) for both the tree and the animal. Because of this, Jacob Gerstner, the Zululand missionary-botanist, gave this tree its appropriate botanical name (1972, vol. 1: 465).

It is not exactly true to say 'the Zulus use the same name . . . for both the tree and the animal.' It would be more accurate to say the class 3 noun *umvubu*, referring to the tree, is derived from the class 9 noun *imvubu*, referring to the animal.[26] And then we must note that north of the uThukela River, the animal is known as *imboma*, with the noun *imvubu* referring more specifically to a whip made of hippo hide. I would suggest that Gerstner gave the specific name *hippopotami* because he was aware of the derivation of the tree name *umvubu*.

Returning to the topic of fruit reflected in Zulu plant names, we can note **uqhamuqwingi** (< *qhamu* + *qwingi* (that which appears and it is ripe)), a name Doke and Vilakazi (1958: 696) give for an unidentified bush bearing small, red, edible berries.

A tuber, rather than a fruit, is described in **iqanda-lenkuku** (chicken's egg), a name Doke and Vilakazi (1958: 688) give to an unidentified species of sweet potato.

Very thin twigs are referred to in the Zulu name **ubambo-lweyoka** (snake's ribs), the name for the Sand Peawood (*Craibia zimmermannii*).[27]

A completely different physical characteristic is 'exploding with black spores'. I am unable to trace elsewhere Doke and Vilakazi's *Lycoperdinea* fungus, which they say is "filled with powder of black spores and bursting when ripe" (1958: 284). They give the English common name Snuff-Box and this

---

26. This kind of derivation, involving the moving of a noun stem from one noun class to another, was discussed in detail in Chapter 3.
27. More correctly, says Beaumont (personal communication), the name refers to "the very thin rachis and petiole which is apparent in their leaves".

metaphor for black spores was either coined independently, or was influenced by the three Zulu names that Doke and Vilakazi give for this plant, all using the word *ugwayi* (snuff, tobacco): **ugwayikakholo** and **ugwayikanhloyile** both mean 'yellow-billed kite's snuff' and **ugwayikathekwane** means 'hamerkop's snuff'.

While not a Zulu plant name, it is interesting to note the Shona name for the Sickle Bush (*Dichrostachys cinerea*). Coates Palgrave tells us that "the Shona name, *mupangara*, means 'tassels for the chief's hat', and is a picturesque reference to the flowers" (1977: 254).

The 'elephant's ear': Metaphors from the natural world

A large number of Zulu plant names are similar in structure and thought to English plant names[28] such as Elephant's Foot (**Testudinaria elephantipes*), Cat's-Foot (*Antennaria dioica*), Bullock's-Heart (*Annona reticulata*), Foxtail (any grass of the genus *Alopecurus*) and Goatsbeard (*Tragopogon pratensis* and *Aruncus sylvester*).

Tongues are particularly associated with cattle and buffalo, and tongue-based names usually refer to leaf shape, as in the following two Zulu names: **ulimilwenkomo** (cow's tongue), the Pink-and-White Gerbera (*Gerbera ambigua*) with long tongue-shaped leaves, very hairy above (this name also refers to *Melanthera scandens*); and **ulimi-lwenyathi** (buffalo's tongue), the Leg-Ripper or Thorny Rope (*Smilax anceps*).

Both **ulimilwenkomo** and **ulimilwenyathi** are names for the similarly named Buffalo-Tongue Berkheya (*Berkheya setifera*). The leaves of this plant have coarse, straw-coloured bristles on the upper surface.

Ears form the basis for the following four names:

- **indlebeyemvu** (sheep's ears): the Sheep's Ears Everlasting (*Helichrysum appendiculatum*). Again, both English and Zulu names coincide in referring to the appearance of this plant and Pooley tells us the "stems [are] grey, woolly . . . White underside of leaves stripped for fringes and body ornaments in the past" (1998: 82);
- **indlebe-kathekwane** (hamerkop's[29] ear): various *Plantago* species. It is not clear what part of the plant is being referred to;

---

28. Examples from *Collins Dictionary of the English Language*.
29. The bird *Scopus umbretta*. As birds do not have external ears, it is probably the noticeable crest of this bird that is referred to.

- **indlebeyempithi** (bush-buck's ear): the Small Yellow Gerbera (*Gerbera piloselloides*); and
- **idlebelendlovu** (elephant's ear): a name given to the Brown Ironwood (*Homalium dentatum*). Boon refers to its "large roundish leaves" (2010: 374). Boon also offers the name **umkhakhasi** for this tree, a name Doke and Vilakazi explain (1958: 375) as derived from the noun *ikhakhasi* (broad, blade-shaped leaf or article). The name **idlebelendlovu** is also used for the trees Triangle Tops (*Blighia unijugata*), Wild Mulberry (*Trimeria grandifolia*) and Small-Leaved Wild Mulberry (*T. trinervis*).

The next three examples are based on the shape of feet:
- **unyawolwengulube** (pig's foot): the perennial weed Batchelor's Button (*Gomphrena celosioides*);
- **unyawolwendlovu** (elephant's foot): the Wild Yam (*Dioscorea cotinifolia*) with the Afrikaans name *olifantsvoet* (elephant's foot). Both names refer to the shape of the tuber; and
- **unyawolwenkuku** (hen's foot): the Stalk-Flowered Pelargonium (*Pelargonium luridum*). This name appears to refer to the shape of the flower (see illustration on page vi).

Linked to the foot metaphor is that of the footprint, as seen in **isidladlasengwe** (leopard's pawprint), a name for the Coast Climbing Thorn (*Acacia kraussiana*).

Moving to yet another body part, we find the name **amabelejongosi** (breasts of the young girl), referring to the epiphyte *Polystachya ottoniana*.

From breasts to heads, crowns and crests: the ratel (*insele*) is personified in the name **umakhandakansele** (the heads of Mr Ratel), a name for the Common Pineapple Flower (*Eucomis autumnalis*) with its several large flower heads. From the heads of a ratel to the head of a cat: **ikhandalempaka** ([mythical] cat's head)[30] is the name of the Common Coca Tree (*Erythroxylum emarginatum*).

The large crown-like inflorescence of the Giant Candelabra Flower (*Brunsvigia grandiflora*) is reflected in the Zulu name **umqhelewenkunzi** (bull's crown) as well as in the specific name.

References to eyes can also be found in Zulu plant names. Doke and Vilakazi (1958: 355) provide the following examples: **isolemamba** (mamba's eye)

---

30. An *impaka* is a kind of wild cat believed to be especially bred by *abathakathi* (witches) in order to assist them to carry out their evil deeds.

refers to both the tree *Cassinopsis tinifolia* as well as the shrub *Syncolostemon densiflorus*. The latter is also known as **isolenyoni** (bird's eye). The Sow-Thistle *Sonchus ecklonianus* has the Zulu name **isolendlovu** (elephant's eye) and the shrub *Jasminum multiflorum* is called **isolenkosazane** (maiden's eye) in Zulu.

Moving from the eyes at the front to the tail at the back, we find **umsila-wengwe** (leopard's tail), a name given by Doke and Vilakazi for a "species of shrub, *Gnidia kraussii*, used medicinally for stomach complaints" (1958: 755). This is presumably Hutchings' *Gnidia kraussiana* var. *kraussiana* (1996: 211). Pooley (1998: 292) gives **umsilawenge** [*sic*] as both the Zulu and Swazi names of *Gnidia kraussiana*, but no other information that might explain 'leopard's tail' (see Frontispiece).

Two more 'tail-names' are given in Doke and Vilakazi, these being **ishoba-lehhashi** (horse's tail) for "species of grasses, *Equisetum*[31] *ramosissimum*, etc." and **ishoba-lenyathi** (buffalo's tail) for a "species of fern used as a love-charm, *Pellaea viridis*" (1958: 743).

Genitalia

Genitalia feature strongly in Zulu plant names of the 'elephant's ear' type. The noun *igolo* refers to "abdominal aperture, e.g. the vulva of the female, and the anus" (Doke and Vilakazi 1958: 254, who say this term should be avoided). We find this reference in:

- **igololembuzi** (goat's vulva/anus): the name of the Weeping Bride's Bush (*Pavetta lanceolata*);
- **igololenkawu** (monkey's vulva/anus): the Dune Soap-Berry (*Deinbollia oblongifolia*); and in
- **igololenkonyane** (calf's anus): Doke and Vilakazi say this refers to a "species of fruit-bearing vine, *Cissus connivens*, found along rivers in the mist-belt areas" (1958: 254).

Doke and Vilakazi say of the word *umsunu* that it is a vulgar term for the vagina and that "[t]his term is used, esp. by Native doctors, in certain descriptive names for medicinal plants" (1958: 770). They give as examples **umsunu-wembuzi**

---

31. Which I am guessing means 'horse's tail'.

(goat's vagina) for *Pavetta lanceolata*,[32] and **umsunu-wengane** (child's vagina) for "spp. of plants *Vangueria edulis*, *V. latifolia*".

A similar word to *umsunu* is *inhlunu*, which Doke and Vilakazi regard as an 'obscene' term for the vagina rather than merely a vulgar one. Their explanation, however, starts off with exactly the same words: "[t]his term is used, esp. by Native doctors, in certain descriptive names for medicinal plants". Their examples of plant names based on *inhlunu* are:

- **inhlunu-yembuzi** (goat's vagina): the "Christmas tree with white bloom, *Pavetta caffra* and *P. lanceolata*";
- **inhlunu-yewabayi** (raven's vagina): "a large forest tree"; and
- **inhlunu-yomntwana** (child's vagina): "*Vangueria latifolia* tree, with edible fruit" (Doke and Vilakazi 1958: 339).

We noted previously that the Weeping Bride's Bush could be referred to by Zulu *izinyanga* as **igololembuzi**, **umsunu-wembuzi** and **inhlunu-yembuzi**. Boon (2010: 564) records only **igololembuzi**, while Hutchings records no names at all for *Pavetta lanceolata*. Neither *Vangueria edulis* nor *Vangueria latifolia* appear in either Boon or Hutchings, so I am unable to say how they have treated the names **umsunu-wengane** and **inhlunu-yomntwana**. Note that Ndela Ntshangase says that names based on *igolo*, *umsunu* and *inhlunu* are "not for public use, or for children's ears".[33]

Boon (2010: 140) gives **isikhabamkhombe** as one of the Zulu names for the African Wattle (*Peltophorum africanum*). The name is derived from *isikhaba* (animal penis) and *umkhombe* (white rhinoceros). A semantically similar name is **umthondo-wemfene** (baboon's penis) given by Hutchings as one of the names of the Jeukui (*Schizobasis intricata*).

Surprisingly enough, Doke and Vilakazi have no examples of plant names based on *umthondo* (penis; they do not have Hutchings' **umthondo-wemfene**) but they do offer two names for various *Eucomis* lilies based on the word *umnqundu* (animal penis), namely **umnqundu-wenkunzi** (bull's penis) and **umnqundu-wenyathi** (buffalo's penis). These names are apparently used for *Eucomis regia*, *E. undulata* and *E. axyrioides*. Pooley (1998) does not recognise

---

32. Hutchings (1996: 240) gives **umsunu wembuzi** as the name of *Nuxia floribunda*.
33. Personal communication.

any of these botanical names. Hutchings (1996: 42) gives *Eucomis regia* and *Eucomis undulata* as synonyms for two subspecies of *Eucomis autumnalis*, but does not mention any *umnqundu*-based names.

The Natal Bush Milkwood (*Vitellariopsis marginata*) shows an unusual combination of body parts in its Zulu name: **umnqambomabele** (ram's penis + breasts).

Zulu, Afrikaans and Latin disagree about which metaphor is appropriate for the Dune Morning Glory. Pooley (1998: 422) gives **isendelengulube** (testicle of pig) and **ubulilibengulube** (sex of pig) as the Zulu names, *strandpatat* (beach potato) as the Afrikaans name and *Ipomoea pes-caprae* (goat's foot) as the botanical name. Both Boon (2010: 362) and Hutchings (1996: 202) give the name **isendelengulube** in the abbreviated form **isendengulube** for the Natal Plane (*Ochna natalitia*).

While on the subject of testicles, we should note **umthekwini**, a Zulu name for the Tongakierie (*Cladostemon kirkii*). The name **umthekwini** is the locative form of *umtheku*, a class 3 noun derived from the class 5 noun *itheku*, meaning both 'bay' or 'lagoon' and 'person or animal with a single testicle'. Both **umthekwini** and Tongakierie refer to the testicle-sized and testicle-shaped fruit, perfectly illustrated in the photo accompanying Boon's text (2010: 116, 117). The plant name **umthekwini** is closely related to eThekwini, the Zulu name for Durban usually translated as 'the place of the bay'. It has long been held that an alternative translation is 'the place of the single testicle'.[34] Both Boon (2010: 216) and Hutchings (1996: 155) give **umthekwini-omncane** (little thekwini) for *Commiphora zanzibarica*, with Boon calling this tree the Pendent-Fruit Corkwood and Hutchings calling it the Pongola Corkwood.[35] The fruit in Boon's picture does indeed look like a smaller version of the fruit of the Tongakierie.

Zulu is not the only language that uses testicular metaphors in the naming of plants. The plant *Solanum candidum*, found from Mexico to Peru, produces small, round fruit that when cultivated are known as 'naranjillas'. Heiser (1985: 68) says of this plant that it is most commonly known as *huevo de gato* (testicle of the cat).

---

34. See Koopman (2002: 157–63) for a summary of the long-standing debate on this and Koopman (2009) for an extensive re-evaluation of the meaning of 'eThekwini'.

35. Not to be outdone, Moll (1981: 464) calls this tree the Zanzibar Commiphora. He gives the Zulu name as **umThekweni-omncane** [*sic*].

Shape, size, texture

Various aspects of the tree are referred to here: the overall size (large, small), growth patterns (twisted, stunted), and the size, shape and texture of leaves and fruit.

Starting with size, we find names that range from the large, as in the Belhambra Tree (*Phytolacca dioica*), with its Zulu name **umzimuka** derived from the verb *zimuka* (grow stout, large), to the smaller, more stunted type of growth, as in the Porkbush (*Portulacaria afra*), its Zulu name **isidondwane** derived from the ideophone *dóndo* (of being stunted).

A straight stem is featured in **umkhombazulu** (what points to the sky), a Zulu name for the Cape-Teak Bitterberry (*Strychnos decussata*). The fruit of the Mountain Nettle (*Obetia tenax*), described by Pooley as "very small, thin, in dense clusters" (1993: 80), gives rise to Zulu **impongozembe** from *impongoza* (long limp rope-like mass) + *embe* (other). This name could also be interpreted as derived from *impongo* (he-goat) + *izembe* (axe), but it would be difficult to relate this interpretation to the plant.

Another name that refers to fruit is **inhlamvumabele** (grain of sorghum). Doke and Vilakazi say that this refers to "certain sp. of shrubs, **Maesa rufescens*, etc, bearing white berries" (1958: 318). Hutchings (1996: 227) gives *Maesa rufescens* as a synonym for *Maesa lanceolata*, of which Boon says "fruit cream, round . . . in dense clusters" (2010: 444). Certainly the fruit in his accompanying picture bears a strong resemblance to sorghum.

Also from Boon (2010: 448) is the name **umthungulu** for the Milkpear (*Inhambanella henriquesii*). Doke and Vilakazi (1958: 807) say this word is from Ur-Bantu *tuŋgula* (be globe-shaped), a clear reference to the fruit and also an indication of a very old plant name.

Boon (2010: 336) gives **isibinda** as one of a number of names for the Cat-Thorn (*Scutia myrtina*).[36] According to Doke and Vilakazi (1958: 78) this name is derived from the verb *binda* (choke, get stuck in the throat). Boon explains this name, and the Afrikaans name for the tree *droog-my-keel* (dry my throat), as "fruit astringent, dries out [the] mouth" (2010: 336).

---

36. We met the other names earlier in this chapter under the heading 'Thorns': **usondela** (come closer), **usondeza** (bring closer), **usondelangange** (come closer so I can kiss you) and **umbambangwe** (what catches the leopard).

While still on the subject of fruits, we could note that Doke and Vilakazi give **ukhamginqi** (pluck and devour) as the name for a "sp. of large fruit tree, *Dovyalis rhamnoides*" (1958: 378). Boon (2010: 372) says of this tree that its tasty fruit is eaten by humans and used to make jelly preserve.

Abundance of fruits is suggested in the name **uguguvama**. Doke and Vilakazi translate it as "a precious thing found abundantly" and give it as the name of a "species of shrub with edible berries, *Lantanea salvifolia* [=*Lantana salviifolia*]" (1958: 274). Hutchings (1996: 262) helps us to see that this is *Lantana rugosa* and we read in Pooley that this plant, with English common names Bird's Brandy and Wild Grassland Lantana, has "juicy purple fruit clustered in bracts, which are eaten by people, monkeys and birds (1998: 422).

The thin, wiry growth of the Mountain Laburnum (*Calpurnia aurea*) is suggested in **insiphane-enkulu** (< *insiphana* (thin, strong, wiry person) < *umsipha* (muscle, sinew) + *enkulu* (large)) in contrast to the **insiphane-encane** (small insiphane) that Doke and Vilakazi (1958: 759) tell us refers to the Forest Indigo *Indigofera natalensis*, a slender, delicate tree or shrub with thin stems.

The English name and Zulu names share a common reference to the cone-shaped knobs on the stem of the Small Knobwood (*Zanthoxylum capense*). Pooley (1993: 184) gives four Zulu names: (1) **umnungumabele** (**umnungu** (knobwood tree) + **amabele** (breasts)); (2) **umnungwane omncane** (small little umnungu); (3) **amabelentombi** (maiden's breasts); and (4) **amabelezintshingezi** [*sic*], apparently a corrupted form of **amabelejongosi** (young maid's breasts). Zukulu et al., in discussing the Pondo name *umlungumabele* for this tree, add an interesting extra feature to the name:

> Before colonial times the name of this tree was called *uMnugumabele*, meaning 'thorns that resemble breasts'. After the arrival of European women in the country the name was changed to *umlungumabele*, meaning 'white woman's breasts' because the large woody cone-shaped knobs on the bark resemble the breasts of a woman wearing a brassiere (2012: 16).

If this is true, which I doubt, then it would certainly be an unusual example of Euro-Western influence on indigenous botanical nomenclature.

Seed pods that can be used as rattles are the basis of the names of the next two species. The Snuff-Box Tree (*Oncoba spinosa*), with "fruit shells used for rattles for dancers, for snuff boxes and for protective penis covers"

Plate 1: The Snake Lily (*Scadoxus puniceus*) has the Zulu name **idumbe likanhloyile** (the tuber of the yellow-billed kite), referring to the fact that both plant and bird appear at the same time of the year: towards the end of July. This is the Zulu month of *uNcwaba*, also known as *uNhloyile* (yellow-billed kite month) (drawing by Angela Beaumont).

Just as plants are found in certain habitats, so too are plant names. Here are three such habitats: in dictionaries, on name plaques on trees and on labels in a 'muti market'.

Plate 2: A name plaque on the trunk of a Red Milkwood, showing names in Latin, English, Ndebele and Shona. The photograph was taken by the author on a visit to Victoria Falls in September 2014. Opposite is an extract from Wild (1952: 104) showing the same names, together with some other African vernacular names.

Plate 3: A name plaque on a Waterberry growing at Victoria Falls. Opposite is an extract from Wild (1952: 129) showing the same names. The Shona name *Muhute* on the plaque is presumably a variant of the name *muKute-shenje* recorded by Wild. Van Wyk et al. (2011: 640) give the following Shona names for this tree: "hute (*fruit*), muisu mukute, mukute-shenje, muwototo, shenje (*fruit*)".

at the apex of the short branchlets, ferruginously hairy, particularly in bud.
Fruit not seen.

Mimusops zeyheri Sond. (Sapot.)
> Red milkwood, *umBumbulu*, *uChirinji*, *uTunzi*, *muShaplha*, *muCheningi*, *muChechete*

A widely distributed medium-sized tree with milky latex and scaly bark. Young twigs and leaves closely rufous tomentose. Leaves alternate, petiolate, oblong lanceolate, obtusely acuminate, leathery, glabrous. Flowers on longish pedicels, two or three together in the axils, rufous tomentose. Fruit an elliptical yellow drupe, about 1in. long, edible.

Momordica balsamina L. (Cucurbit.)
> *nGaka*

A slender climbing tendrilled herb with alternate leaves. Leaves

Plate 4: Hiram Wild (1952: 104) *Mimusops zeyheri* entry showing Ndebele name **umbumbulu** and other names for the same tree.

Plate 5: A plant name habitat: handwritten tabs inserted into bags of chopped roots, corms, tubers, etc., photographed by Chris Ellis at the Warwick Avenue 'muti market' in Durban.

dure, hard, dark brown, shining, up to 6in. long and 3in. in diam.

Syzygium cordatum Hochst. (Myrt.)
> Water boom, Water berries, umDoni, Garnumkela, umSwi, muNony-anansi, muToya, muKute-shenje

A common and widely distributed tree, with a compact, rounded crown, common along river sides. Leaves opposite, cordate, oblong, sessile or subsessile, rounded at the apex, glabrous, entire, leathery, dark green. Flowers white or yellowish in terminal panicles. Fruit oval, fleshy, purple, the size of a small plum, edible.

Syzygium gerrardii (Harv.) Hochst. (Myrt.)

Plate 6: Extract from Wild (1952: 129): entry for *Syzygium cordatum*, showing Ndebele name **umdoni**, and other names for this tree.

Plate 7: Baskets woven from strips of palm leaf.

Plate 8: *Hyphaene coriacea*, the Lala Palm (Zulu **ilala**). The Lala Palm and the Raphia Palm are both used to make traditional baskets. Their Zulu names, **ilala** and **umvuma** respectively, link the trees, the baskets and Zulu culture in an unusual way, with *ilala* at one time meaning "one skilled in craftwork" and *umvuma* derived from the verb *vuma*, with an earlier meaning of "turn out well", a reference to beautifully made craftwork (see pages 16 and 227) (photo courtesy of Richard Boon).

Plates 9–10: An *umncedo* (penis sheath) made of strips of the Wild Banana (*Strelitzia nicolai*) is known as an *ingceba* or an *inkamanga*, the latter word also being a name for various species of Strelitzia.

Plate 11: The Fever Tree (*Acacia xanthophloea*) is known (among other names) as **umkhanyakude** (what shines from afar), a reference to the light yellow-green bark (see page 133) (painting by the author).

Plate 12: The Paperbark (*Acacia sieberiana*) is known as both **umkhamba** and **umkhambathi** (from *umkhamba* + *(umu)thi* 'tree') (see page 64).

Plates 13–14: The Cape Fig (*Ficus sur*) has the Zulu name **umkhiwane**, while its fruits are known as *amakhiwane* (singular *ikhiwane*) (see page 50).

Plate 15: Beautiful as *Clivia miniata* is, its Zulu name **umayime** refers to its power to stop evil in its tracks (see pages 57, 111, 167 and 189).

Plate 16: The photo shows the distinctive leaf pattern of *Harpephyllum caffrum* throwing an equally distinctive shadow on the wall behind. The link between this tree – **umgwenya** – and the crocodile (*ingwenya*) is discussed on page 235.

Plate 17: The large pendulous fruits of *Kigelia africana* provide both the English name Sausage Tree as well as the Zulu name **amabelendlovu** (elephant breasts).

Plate 18: *Haworthia limifolia*, with the Zulu name **umathithibala** (what renders helpless), is often planted around the edge of a homestead to render potential evil-doers helpless (see page 191).

Plate 19: The Zulu name **ubani** (lightning) is used for the Agapanthus. It is frequently used as a protective charm against lightning (see page 201).

Plate 20: Both the Zulu Loquat (*Oxyanthus latifolius*) and the Canary Sand-Berry (*Suregada zanzibariensis*) have the Zulu name **umdlankawu** (what the monkey eats) (see pages 115 and 150). This photo of a vervet monkey eating loquats was taken in the author's garden.

Plate 21: The Assegai Tree (*Curtisia dentata*) has the Zulu name **umlahleni** (throw him away), a reference to its use in witchcraft preparation (see page 200).

Plate 22: The Buffalo-Thorn (*Ziziphus mucronata*) has the name **umlahlankosi** (lay the chief to rest), a reference to its use in the *ukubuyisa* ceremony held a year after a person's death (see page 220).

Plate 23: The Cycad has the Zulu names **isigqiki-somkhovu** and **isidwaba-somkhovu**, both linking this plant to the most dreaded of witches' familiars, the *umkhovu* (zombie) (see pages 197–8).

Plate 24: The **umthongwane** or Snuff-Box Tree (*Oncoba spinosa*) also has the Zulu name **isingongongo**, an onomatopoeiac name referring to the use of the fruit in rattles, like the one pictured (see page 143).

(Pooley 1993: 326), has the Zulu name **isingongongo** (Ngo! Ngo! Ngo!), an onomatopoeic name referring to rattling (see Plate 24). English, Afrikaans, Latin and Zulu names all make the same point in the shrub *Crotolaria capensis*. The Latin word *Crotolaria* means 'castanets', referring to the sound of seeds in the pod. The English name is Cape <u>Rattle-Pod</u>, the Afrikaans name *Kaapse klapperpeul* and the Zulu name is **ubukheshezane** (< *khesheza* (rattle as seeds in a dry pod)).

Quite why the Wild Pride-of-India (*Galpinia transvaalica*) should be regarded as leafier than other trees is not clear, but the Zulu name **umphisamakhasi** means 'what gives leaves'. Another leaf reference is found in **unwele oluncane** (small hair), the name of the Brown Gonna (*Passerina filiformis*)[37] in reference to its "very narrow, thread-like leaves" (Pooley 1993: 342). This could also perhaps be a reference to the use of the tough fibrous bark of this tree as binding twine. The same metaphor is used in **inwele**, the Zulu name of the Clubmoss (*Lycopodium clavatum*). Hutchings says "the plant is named for its resemblance to human hair" (1996: 5). Hair, or rather <u>fur</u>, is the underlying metaphor in **insindeboya**, a name that Doke and Vilakazi (1958: 599) assign to the grass species **Trichopteryx simplex*. The name is a compound of **insinde**, which Doke and Vilakazi give for the "bluish-red veld grass, *Anthistiria ciliata* and *A. australis*" (1958: 599) and *uboya* (fur).

Still on the topic of leaves, we note **ukhasikhulu** (big leaf), given by Doke and Vilakazi as "sp. of large-leaved shrub" (1958: 383). Neither Pooley nor Hutchings recognises the name.

Plume-like growths are suggested in the next two names: **umbhongothi** (< *umbhongo* (ostrich plume) + (*umu*)*thi* (tree)), the Sausage Tree (*Kigelia africana*); and **intshakaza** (heavy head of corn, heavy bamboo top), a name for both the Cape Sumach (**Colpoon compressum*) and the Transvaal Sumach (*Osyris lanceolata*).

Along the same lines is the name **umkhakhayi**, which Doke and Vilakazi say refers to "[a] large timber tree, *Mimusops caffra*, [which] has flowers resembling the tuft on a duiker's head" (1958: 375) (see illustration on page 144). They say that the name is derived from the noun *u(lu)khakhayi* (crown of head). Boon (2010: 452) does not recognise the name, nor does he give any indication of such a tuft-like flower.

---

37. *Filiformis*, says Pooley (1993: 342), means "thread-like" and is also a reference to the narrow leaves.

Doke and Vilakazi (1958: 375) say that **umkhakhayi**, a name for *Mimusops caffra*, is derived from the noun *ukhakhayi* (crown of head), and that the tree "has flowers resembling the tuft on a duiker's head" (see page 143) (drawing by Angela Beaumont).

Characteristic habitat

Plant names under this heading give some clue as to where the plant is customarily found: in marshes, open grassland, stony ground, and so on. These names can be roughly divided into two groups: those that refer to rocks, cliffs and hillsides, and those that refer to wetlands.

Rocks, cliffs and hillsides

The Cork Bush (*Mundulea sericea*), found on rocky hillsides, has the Zulu name **umamentabeni** (standing on the hill). The name consists of [*u* + *ma* (characteristic) + *ma* (stand) + *entabeni* (on the hill)]. Similar in meaning, but using the metaphor of a human head with hair on the top, is **unwelelwentaba** (hair of the mountain), the name of the Mountain Cypress (*Widdringtonia nodiflora*) that is found "growing in large groups on exposed mountain slopes" (Pooley 1993: 48).

Ubambematsheni (what catches hold of the rocks) is a name used for the Large-Leaf <u>Rock</u> Fig (*Ficus abutilifolia*) as well as the Small-Leaved Coca Tree (*Erythroxylum delagoense*) that grows on sandstone outcrops; and the

Common Coca Tree (*E. emarginatum*) and Forest Coca Tree (*E. pictum*), both of which grow on rocky hillsides and stone outcrops.

Hutchings (1996: 202) gives **umilamatsheni** (what grows among rocks) as a Zulu name for the Cape Plane (*Ochna natalitia*), which Boon says is "often [found] among rocks" (2010: 362).

Also pointing to rocky habitat is **insema-yamatshe** (< **insema** (various species of ground Euphorbia) + *yamatshe* (of the rocks)) that Doke and Vilakazi give as "species of small shrub, *Pachypodium saundersii*, growing on dolerite rock" (1958: 598). Pooley recognises **insema-yamatshe** as one of two names for *Pachypodium saundersii* (the Kudu Lily) and says "succulent shrub . . . in dry rocky areas" (1998: 166).

Based on the locative *eweni* (on the cliff) are the names of the three trees:

- Forest Fever-Berry (*Croton sylvaticus*) with the name **ugibeleweni** (climbing up the cliff);
- Wild Maple (*Seemannaralia gerrardii*) known as **umaweni** (on cliffs); and
- Krantz Aloe, Afrikaans *kransaalwyn* (*Aloe arborescens*), which has the Xhosa name *unomaweni* (on the cliffs). Doke and Vilakazi (1958: 584) give the same word, **unomaweni**, as the Zulu name for *Aloe nitens*.

Doke and Vilakazi give **igqamamaweni** (what stands above the precipices) as the name of *Begonia dregei*, a "species of edible flowering plant" (1958: 261). **Ugibeleweni** (mounted on a rock) is also a name for the Hanging Wild Cactus (*Rhipsalis baccifera*), a "pendulous, epiphytic succulent . . . [found] . . . on rocks in evergreen forests" (Pooley 1998: 158). Also found "on cliff faces in kloof forests" (484) is the perennial herb *Streptocarpus prolixus* with the Zulu name **isikhwalisamatshe** (**isikhwali** of the rocks), where the word **isikhwali** refers to various species of tuberous plants and climbers.

The Yellow Starwort (*Senecio bupleuroides*), a veld herb growing among stones, has the name **isiqandamatshana** (what pounds little stones) and the Sticky-Leaved Rubia (*Rubia cordifolia*) is known as **intwalalubombo** (that which climbs the ridge).

Wetlands habitat

Turning now to plant names that indicate a wetlands habitat, we find the Wild Frangipani (*Voacanga thouarsii*) that requires swampy conditions with the Zulu name **umhlambamanzi** (what swims in water). The Wild Pomegranate

(*Burchellia bubalina*), which needs similar wet conditions, is known as **umvuthwemfuleni** (what ripens in the river).

Boon (2010: 538) gives **umfula** (river) as one of the Zulu names of the Matumi (*Breonadia salicina*), an almost certain reference to the fact that this tree is found on watercourses in the bushveld.

Pooley (1998: 252) does not give a distribution map for the alien Yellow Mexican Poppy (*Argemone mexicana*), but its Zulu name suggests a very limited locality: **ugudluthukela** (what grows along the Tugela River).

Umnyamanzi (< **umunya** (sp. of evil-smelling tree) + *amanzi* (water)) is the name of the Large-Leaved Onionwood (*Cassipourea gummiflua*) found in swamp forest. A very similar name is **umngamanzi** (< **umunga** (acacia) + *amanzi* (water)), which refers to the Ankle Thorn (*Acacia robusta*).

Doke and Vilakazi and Boon concur that **amanzana** (small amount of water) is a Zulu name for *Eugenia natalitia*, the Forest Myrtle (Doke and Vilakazi: "Natal myrtle bush" (1958: 485)), but Boon says this tree is found "in forest & margins & among rocks in grassland" (2010: 416) unlike the related species *Eugenia simii* (River Myrtle) found on watercourses and for which Boon gives no Zulu names at all. An unidentified tree is reflected in **ilalamanzini** (species of tree growing near water; Doke and Vilakazi 1958: 444).

Two more from Doke and Vilakazi (1958: 109, 120) are **umchaphamanzi** (the water-lapper), an unidentified species of fruit tree growing near water, and **umcika-manzi** (< *ciki* (of fullness) + *amanzi*, a species of tree or shrub, **Conopharyngia ventricosa*, growing along water.

The Common Buttercup (*Ranunculus multifidus*) according to Pooley (1998: 252) grows in damp ground nears streams and marshes. This is reflected in its Zulu name **uxhaphozi**,[38] which means simply 'marsh' or 'bog'.

The Jumping-Bean Tree (*Spirostachys africana*) is probably better known to English- and Afrikaans-speaking South Africans as the Tambootie or Tambotie. This name is adopted from Zulu **umthombothi**, a compound of *umthombo* (spring, fountain) and (*umu*)*thi* (tree). The tree is normally associated with groundwater. As we saw in earlier discussion, it could be that *umthombo* is an obsolete Zulu word meaning 'stone', in reference to the hard, heavy wood.

---

38. Doke and Vilakazi refer to this as "species of marsh herb, **Ranunculus pinnatus*" (1958: 917). Hutchings (1996: 110) confirms *Ranunculus pinnatus* as an earlier synonym of *Ranunculus multifidus*.

Habitat reference via animal reference is the basis of **umvubu**, the Zulu name of the Swamp Fig or Hippo Fig (*Ficus trichopoda*) discussed in some detail on page 135.

Found in marshy areas (i.e., in the environment of crab and frog) is the Pink Marsh Dissotis (*Dissotis canescens*) with the two Zulu names **imfeyenkala** (sweet-reed of the crab) and **imfeyesele** (sweet-reed of the frog). In similar vein is the name **inkwindi** (mussel) for the Sea-Bean (*Entada rheedii*), which Boon tells us grows "in swamp and coastal forest and estuary margins" (2010: 196).

Two plant names that indicate neither rocky hillsides, nor wetland are **umadotsheni** (based on the locative plural of *idobo* (low-lying valley land)), a name for the Yellow Everlasting Flower (*Helichrysum cooperi*) and **umhlalanyoni** (where the birds sit), a reference to the Natal Mistletoe (**Tapinanthus natalitius*) that grows high up in trees; i.e., 'where the birds sit'.

Doke and Vilakazi (1958: 204) tell us that **imfenyane** is a name for *Senecio rhyncholaenus*, a composite growing in sandy, watery places. The name is derived from *ifenya* (soft, moist alluvial soil). Incidentally, iFenya is an old Zulu name for the flatlands between the Bay and The Bluff in Durban.

Perhaps one of the most 'localised' of Zulu habitat names is **umnyankomo** (where the cow defecates), which Doke and Vilakazi say refers to a "species of fine, soft grass, *Eleusine indica*, common outside cattle kraals" (1958: 621).

Also rather localised in the sense of 'found amongst thorns' is *Begonia sutherlandii* for which Pooley (1998: 68) gives **uqamamaweni**. Doke and Vilakazi (1958: 696) have **uqhamamaveni** (< *qhama* + loc. *ameva*; i.e., what is prominent among thorns).

Phrasal habitat names

Finally, under the heading 'Habitat names' we should not forget the hundreds of plants that are compounds. The first part would equate to a genus name in the Linnaean system and the second indicates a specific habitat. The following sets are typical:

- **ihlamvu**: *Gloriosa superba*, *Callilepis laureola* (and others);
- **ihlamvu lasenhla** (down-country ihlamvu): *Sandersonia aurantiaca*;
- **ihlamvu lasolwandle** (coastal ihlamvu): *Gloriosa superba*;
- **ihlamvu lehlathi** (forest ihlamvu): *Littonia modesta*;
- **ihlamvu elimpofu lasenkangala** (pale open-plain ihlamvu): *Disa stachyoides*; and

- **ihlamvu lentaba** (mountain ihlamvu): *Moraea spathulata* (all from Pooley 1998).
- **idungamuzi**: *Euclea daphnoides* and *Euclea natalensis*;
- **idungamuzi lehlanze** (of the bushveld): *Euclea daphnoides*; and
- **idungamuzi lehlathi** (of the forest): *Scolopia* sp. (all from Doke and Vilakazi 1958: 175).
- **umdubu**: *Combretum* species, *Ficus capreifolia*;
- **umdubu wehlanze** (bushveld umdubu): *Combretum glomeruliflorum*;
- **umdubu wehlathi** (forest umdubu): *Combretum kraussii*);
- **umdubu womfula** (river umdubu): *Ficus capreifolia* (all from Doke and Vilakazi 1958: 169).
- **umsishane**: sp. of tree, *Olea laurifolia*;
- **umsishane wehlathi** (of the forest): *Olea laurifolia*; and
- **umsishane wehlanze** (of the bushveld): *Schrebera saundersiae* (all from Doke and Vilakazi 1958: 760).

Flora and fauna relationships

Under this heading we look at names that show how animal and bird life utilise trees.[39] In some cases the tree provides shelter and a roosting place (marked by *hlala* (sit, live), *lala* (lie) and *khwela* (climb up into). In other cases, the tree provides food, marked by the verb *dla* (to eat).

Palmer and Pitman (1972 vol. 1: 161ff) point out that trees provide <u>fruit</u>, mainly for birds and mammals; <u>flowers and nectar</u>, mainly for birds and insects; <u>leaves for browsing</u>, mainly for mammals; and <u>home and shelter</u> for a wide variety of animal life. These relationships between plant and animal are frequently expressed in Zulu plant names.

I shall take 'home and shelter' first. Those Zulu names indicating a tree in which, or in the vicinity of which, a particular animal can often be found include the following:

- **umhlalankomo** (where the cattle sit): Doke and Vilakazi give this as "species of plant with edible fruits, dark purple, from which wine is made" (1958: 315);

---

39. I have found only the names of trees to have this kind of underlying meaning, not the names of herbs, grasses and smaller plants.

- **umhlalampofu** (where the eland lives): Tree Aloe (*Aloe barberae*);
- **umhlalampunzi** (where the duiker lives): White Raisin (*Grewia bicolor*) and <u>Duiker</u>berry (**Sapium integerrimum*);
- **umlalandlovana** (where the little elephant lies): Natal Laburnum (*Calpurnia aurea*);
- **ilalanyathi** (where the buffalo lies): Septee Tree (*Cordia caffra*);
- **isalanyathi** (where the buffalo stays): Puzzle Bush (*Ehretia rigida*);
- **umlalampisi** (where the hyena sleeps): Stink Shepherd's Tree (*Boscia foetida*);
- **umlalampunzi** (where the duiker lies): Giant Raisin (*Grewia hexamita*) and Silver Raisin (*Grewia monticola*);
- **isikhwelamfene** (where the baboon climbs): Water Ironplum (*Drypetes arguta*) and Pigeonwood (*Trema orientalis*).

Palmer and Pitman provide a great deal of detail on the various species of tree that provide fruit favoured by animals. They introduce their section on 'Fruits and animals' with the words:

Fruits provide an even more direct link between birds, and often mammals, and trees, than insects. It is safe to say that all the 200 and more species with fruits that are eaten by humans are likewise used by birds, or monkeys, baboons, elephant and warthogs, and sometimes by a wide range of other mammals as well, while other fruits, unpalatable to people, are also eaten (1972 vol. 1: 171).

For a considerable number of trees, they give specific detail, as for example:

The jakkalsbessie, *Diospyros mespiliformis*, and the nyala tree *Xanthocercis zambesiaca*, always attract large numbers of birds. The Trumpeter, Grey and Yellow-billed Hornbills, Green Pigeons and Brown-headed Parrots may sometimes be seen all feeding on the fruit at the same time. Such birds, monkeys and baboons, knock down and drop as many fruit as they consume, and the fallen fruits, in turn, attract ground birds such as francolin and guineafowl, and nyala, bushbuck and impala. Fruit bats and other nocturnal animals take the fruit at night (1972: 173).

Zulu names that indicate trees favoured as food by various species use the verb *dla* (to eat, which may also be in its passive form *dliwa*). The following are examples:

- **umdlamvubu** (what the hippo eats): Cunningham (1985: 279) gives this as a Zulu name for the grass *Hemarthria altissimia*;
- **indlandlovu** (what the elephant eats): Roundleaved Kiaat (*Pterocarpus rotundifolius*);
- **umdlankawu** (what the monkey eats): Sand Canary-Berry (*Suregada zanzibariensis*) and Zulu Loquat (*Oxyanthus latifolius*) (see Plate 20)
- **umdlampunzi** (what the duiker eats): Jumping-Seed Tree (*Sapium ellipticum, *S. integerrimum*); and
- **umdliwampunzi** (what is eaten by the duiker): White Raisin (*Grewia bicolor*) and Giant Raisin (*Grewia hexamita*). Very close in structure is **umadliwampunzi** with the same meaning, a name Doke and Vilakazi assign to "a species of cress, of the mustard family, *Lepidium capense, *Coronopsis didymos*" (1958: 474).

As Palmer and Pitman point out in the quotation above, antelope species do not necessarily have to reach up for the fruit in trees as much fruit is knocked down by arboreal feeders. They give an interesting example how the pods of *Acacia karroo* are utilised:

> The pods of the sweet-thorn, *Acacia karroo*, and the haak-en-steek, *Acacia tortilis* . . . which are reputed to be particularly rich in protein, are devoured by giraffe, many species of antelope, and by cattle. The seeds of the haak-en-steek, when well swollen but before they are dry, are very popular with baboon which spend long hours perched in their crowns, biting the seeds out and dropping the pods and a good many whole seeds as well, which are picked up by impala and nyala on the ground below. Baboons, according to Howard Kirk, often spread out through a grove of haak-en-steeks, each one with a small attendant group of antelope which follow it each time it moves to another tree (1972 vol. 1: 176).

Returning to the duiker, there are several more Zulu tree names that feature this antelope:

- **umhlolampunzi** (what tests the duiker): given by Doke and Vilakazi for the "bushveld tree *Grewia flava*" (1958: 332);

- **umhlwayampunzi** (what collects the duikers): this Doke and Vilakazi say (1958: 342) refers to the fruit tree *Sapium reticulatum*;
- **umwashampunzi** (what washes the duiker): one of the names given by Boon (2010: 204) for the Horsewood (*Clausena anisata*);[40]
- **umgunguluzampunzi** (what ricochets off the duiker): one of the Zulu names found in Boon (2010: 310) for the Kooboo-Berry (*Mystroxylon aethiopicum*);
- **umnyama-wempunzi** (duiker's omen): Doke and Vilakazi give this as a name for a "species of root medicine" (1958: 619); and
- **umvumampunzi** (duiker palm, < **imvuma** (sp. of palm tree; e.g., *Raphia vinifera*) + *impunzi* (duiker)): *Sapium reticulatum* tree (Doke and Vilakazi, 1958: 842). Hutchings (1996: 174) recognises *Sapium reticulatum* as an earlier synonym of *Sapium integerrimum* and we find a whole herd of 'onomastic duikers' clustered around this botanical name. In addition to **umvumampunzi**, Hutchings gives as **umdlampunzi**, **umhlalampunzi**, **umqathampunzi**, Duiker-Berry and *duiker-bessie*. **Umqathampunzi** is not in Doke and Vilakazi, but it is surely a name along the lines of **isiqathankobe**, which they say is derived from *qatha* (chew) and *izinkobe* (boiled mealies) and which they give as "Fruit of certain Gardenia trees" (1958: 691). This would give **umqathampunzi** an underlying meaning of 'what the duiker chews'.

The duiker undoubtedly features more prominently than any other antelope in Zulu plant names, but it is not the only one. Doke and Vilakazi (1958: 342) record **umhlwambabala** for various species of *Acanthacea* grasses,[41] saying that these are favourite fodder of the bushbuck. The name is derived from **umuhlwa** (species of *Eragrostis* grass) + *imbabala* (bushbuck doe). It seems fairly clear that the name **umhlwampunzi** mentioned above is of the same structure and it seems strange that Doke and Vilakazi translate it as 'what collects the duikers'. Perhaps this is because the name refers to a fruit tree and a translation based on a word for grass would not have been suitable. We note yet a third translation strategy in their entry for **umhlwazimamba**, which they translate as 'mamba climber' and say the name refers to "a species of climber,

---

40. This is the tree we noted earlier as possessing a combined smell of a striped field mouse, horse urine, aniseed and a dirty and perspiring human body, so one wonders how clean the duiker actually is after the washing.
41. Beaumont (personal communication) says there is no such genus for grasses. There is an *Acanthaceae* family, but it does not include grasses.

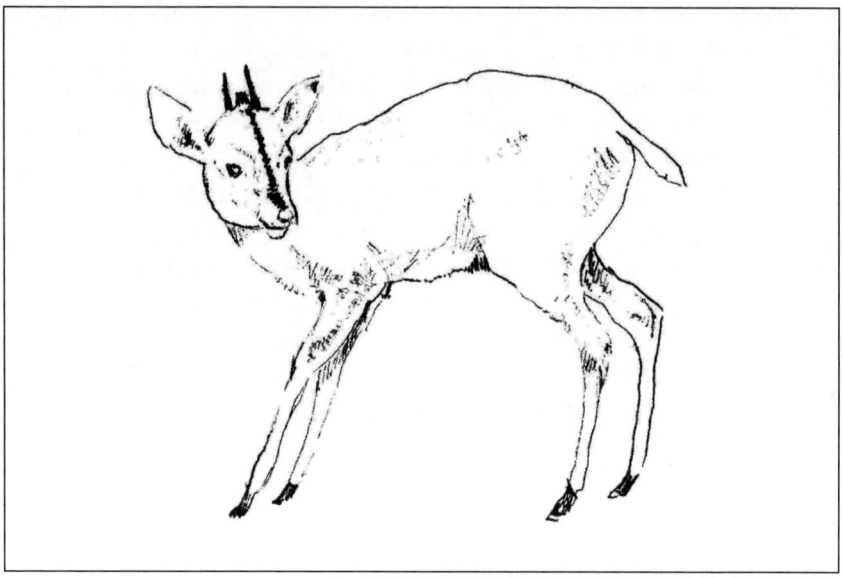

The Zulu name *impunzi* for the duiker occurs in many Zulu plant names (drawing by Adrian Koopman).

the root of which is used for headache" (1958: 342). We must note, however, that this name may well not be from *umhlwa* (grass) + [*i*]*zimamba* (mambas), but rather from *umhlwazi* + [*i*]*mamba* (mamba), where *umhlwazi* could be either 'species of rare tree of the protea veld' or 'species of green striped snake, believed to embody the spirit of a child', both of which meanings are given by Doke and Vilakazi (1958: 342). One would probably need to canvass the opinions of a number of *izinyanga* to see which interpretation is preferred.

The bushbuck doe (*imbabala*) is also singled out in **uphondolwembabala** (bushbuck doe's horn), given by Doke and Vilakazi for a "forest shrub of the Oxyanthus species" (1958: 669). In the plant name **isihlungu-sikankonka**, however, which Doke and Vilakazi (1958: 338) say refers to the thorny shrub *Canthium ciliatum*, it is the male bushbuck (*unkonka*) that is referred to. The name means literally 'bushbuck's poison'. Also referring to a horn is **uphondolwenqama** (ram's horn), which for Doke and Vilakazi is a "species of shrub, prob[ably] *Gasteria glabra*" (1958: 669).

Another herbivore is the wildebeest (Zulu *inkonkoni*) and we find this animal in the name **umhlabankonkoni** (what stabs the wildebeest), referring to the False Forest Spike-Thorn (*Putterlickia verrucosa*). The 'stabbing' refers to its "sharp spines, up to 40 mm long" (Pooley 1993: 272).

We have met the buffalo before in this chapter, in the plant names **ilalanyathi** (where the buffalo lies), **isalanyathi** (where the buffalo stays) and **ulimilwenyathi** (buffalo's tongue). A rather more obscure reference to the buffalo, though, is found in the name **umbulunyathi** (counterfeit buffalo), a name that Doke and Vilakazi assign to a "thornveld scrub, *Osyris abyssinicus*" (1958: 495). Boon (2010: 94) gives *Osyris abyssinica* (not 'abyssinicus') as a synonym for *Osyris lanceolata*, the Rock Tannin-Bush and notes that Zulu, Xhosa and Swati all refer to this plant as a 'counterfeit buffalo'. There is no indication in his description of this shrub to explain this.

A variety of mammals appear in plant names. The elephant features in **ugobandlovu** (what the elephant bends) a name for both the African Mangosteen (*Garcinia livingstonei*) and the Torchwood (*Balanites maughamii*); and in **umsindandlovu** (where the elephant recovers), which refers to the Cork Bush (*Mundulea sericea*). Ndela Ntshangase says that this name refers to a medicine made from this plant, a medicine so powerful that even an elephant will recover from its sickness.[42] Very similar to the last one is **umsindandlovana** (where the little elephant recovers), a name for the Caterpillar Bush (*Ormocarpum trichocarpum*). White rhinoceros and wattle are linked in **isikhabamkhombe** (white rhino clearing), the name for the Weeping Wattle (*Peltophorum africanum*).

The name **umbambangwe:** (what catches the leopard) is given by Boon (2010: 68, 184, 190, 550) for the Thorny Elm (*Chaetachme aristata*), the River Climbing Thorn (*Acacia schweinfurthii*), the Spiny Splinter-Bean (*Adenopodia spicata*) and the Spiny-Gardenia (*Hyperacanthus amoenus*). We also find the leopard in the name **umkhuphulangwe** (what raises the leopard), which for Doke and Vilakazi is a "species of composite herb with purple flowers, used for enemas in feverish conditions" (1958: 417).

Moving to smaller mammals, we find the pig featuring in **igudlangulube** (what the pig rubs up against), a name for the Natal Box (*Buxus natalensis*), as well as in **umpindangulube** (what the pig turns back to), which refers to the Forest Peach (*Rawsonia lucida*). Boon (2010: 108) gives **isilindangulube** (what the pig waits for) as one of the Zulu names of the Cape Wild Quince (*Cryptocarya woodii*). Doke and Vilakazi (1958: 458) say of this tree that it

---

42. Personal communication.

bears a fruit liked by wild pigs. Pooley (1998: 134) gives **intandangulube** (liked by pigs) as one of two Zulu names[43] for the Batchelor's Button (*Gomphrena celosioides*).

We have already met the vervet monkey in the name **umdlankawu** (what the monkey eats). The monkey appears again, more obscurely, in **umzinkawu** (monkey village), a name that Doke and Vilakazi (1958: 895) give for an unidentified species of orchid. This name structure, based on *umuzi* (homestead), is more often found in Zulu river names such as uMzimkhulu (large homestead), uMzimvubu (hippo home) and uMzinyathi (buffalo home).

According to Doke and Vilakazi, a certain "species of Teclea shrub" (1958: 783) is known as **ithambo-ledube** (zebra's bone). The significance of this is obscure. Even more obscure, because a specific plant is not named, is **ithambolenja** (dog's bone), according to Doke and Vilakazi a "species of shrub" (1958: 783).

From dog to cat. Doke and Vilakazi (1958: 597) give the name **insangu-yempaka** to an unidentified species of medicinal plant. The name is a compound of *insangu* (cannabis, marijuana) and *impaka* (a mythical wild cat associated with witchcraft).

Birds

Various species of birds also are linked to verbs like *hlala* (sit, live). The bird species is not identified in **umhlalanyoni** (where the bird sits), the Zulu name for the Tassel Berry (*Antidesma venosum*). The Afrikaans word for this tree, *voëlsitboom* (bird-sit-tree), is clearly a literal translation of the Zulu name.

Again, the general word **inyoni** (bird) is the second half of the compound **umzilanyoni** (what the bird avoids). This is a curious name for the Forest Fever-Berry (*Croton sylvaticus*), for Pooley says of this tree "tree conspicuous when in fruit, attracting many bird species" (1993: 222). There are no questions about the underlying meaning, however, in **utshwalabenyoni** (beer of the bird), the name of the Tree Fuchsia (*Halleria lucida*), for we read that the "nectar-laden flowers attract birds" (Pooley 1993: 448). This name is similar to **utshwala-bentaka**, which Doke and Vilakazi give as "Species of shrub, *Lantana salviaefolia* [= *salvifolia*]" (1958: 824). They translate this name as

---

43. We met the other earlier under the heading 'Elephant's ear metaphors': **unyawo-lwengulube** (pig's foot).

'finch's beer', but as (to my knowledge) finches are seed eaters and not berry eaters, it is more likely that there is some influence from Xhosa here, where the word *intaka* means 'any bird' and not specifically 'finch'.

Back to the verb *hlala*, we find **umhlalajuba** (where the dove sits) a name used for both the Wild Custard-Apple (*Annona senegalensis*) as well as the Forest Fever-Berry (*Croton sylvaticus*).

From dove to raven: **umhlalamagwababa** (where the ravens sit) is a name for the Coastal Goldenleaf (*Bridelia micrantha*) with its "sweet fruit, attractive to birds" (Pooley 1993: 218). Doke and Vilakazi (1958: 624) also assign to this tree the name **umnyelagwababa** (< *nya* + *ela* + *igwababa* (what the raven excretes on)). Boon (2010: 230) does not record this name for *Bridelia micrantha*, but he does give **umhlalamagwababa** for Zulu, *umhlahlamakwaba* and *umhlahlahlungulu* for Xhosa, and *umhlalamagcwababa* and *umhlala-mahubhulu* for Swazi, all names indicating that ravens and crows perch on this tree.

In similar vein is **umhlalankwazi** (where the fish eagle sits), used for both the Ana Tree (*Faidherbia albida*) as well as the Coastal Red Milkwood (*Mimusops caffra*). Doke and Vilakazi (1958: 315) give **umhlalangwazi** (where the noble warrior sits) for an unidentified species of tree, but it seems likely that this is a poorly remembered form of 'umhlalankwazi'. It is possible that the derivation of the name **umhlalankwazi** is not from *hlala* (sit) + *inkwazi* (fish eagle) but from *umhlala* (*Strychnos* tree) + *inkwazi*, to give the meaning 'fish eagle *strychnos*'. This occurs to me as I ponder the two Zulu names Boon (2010: 486) gives for *Strychnos spinosa* (the Green or Spiny Monkey-Orange): **umhlala** and 'umhlalakolontshe'. There is no problem with **umhlala**; all the relevant sources identify it as a name for *Strychnos spinosa*. It is the name 'umhlalakolontshe' which is puzzling. It is not in Doke and Vilakazi, and I would not have expected it to be. Is it perhaps derived from *umhlala* + (*u*)*kholo* (yellow-billed kite) + (*i*)*ntshe* (ostrich)? Ostriches, to my knowledge, feature nowhere else in tree names, but the yellow-billed kite has already appeared in plant names in this book (as in **ugwayikanhloyile**) and it appears in a month name as well.[44]

---

44. The Zulu name for the month that equates to late July to late August in the Western calendar is *uNhloyile* (yellow-billed kite) as this is the time of year the migrating kites return to South Africa (see Plate 1).

The use of the verb *hlaba* (stab, prick) as in **umhlabandlazi** (what stabs the mousebird), Tree Aloe (*Aloe barberae*), seems reasonable for an aloe, but I find no reference to thorns or spines in the similarly based **umhlabahlungulu** (what stabs the crow) for the Tassel Berry (*Antidesma venosum*). Very similar to **umhlabandlazi** is the name **umhlambandlazi**, but in this word the first half is the verb *hlamba* (to wash, to cleanse). The name means 'what washes the mousebird' and it refers to the Forest Elder (*Nuxia floribunda*) for reasons that are not clear to me. Attractive fruit again underlies the name **umdlampangele** (what the guinea fowl eats), the name for the Wild Pride-of-India (*Galpinia transvaalica*).

Insects and reptiles also make their appearance in Zulu tree names and we find **umhlalantethe** (where the locust sits), the name for the Cork Bush (*Mundulea sericea*) and **uhlinzanyoka** (what skins a snake), which refers to the splendidly named Warted Bastard Spike-Thorn (*Putterlickia verrucosa*). The underlying meaning may have something to do with the spines.

According to Doke and Vilakazi if *inyoka* (snake) is added to the name *iphaqa* (an unidentified species of herb) to make **iphaqa-lenyoka**, this now becomes an unidentified "species of fern-leaved weed" (1958: 648). When *inyoka* is added to *ithanga* (pumpkin), on the other hand, to make **ithanga-lenyoka**, this now refers to a "species of wild gourd . . . e.g. *Cucumis myriocarpus*". Doke and Vilakazi explain the name by saying "eaten by snakes" (1958: 785) but somehow this seems as unlikely as finches eating lantana berries.

We end this chapter with snake-related information, given to me by Ndela Ntshangase. It has nothing to do with Zulu plant names, but may be of some interest:

> Snakes, it is believed, cannot reverse. So to kill a snake, one needs to find its hole, and then put over the hole a wooden ring with a razor embedded in it in such a way that a small edge is exposed. The snake will emerge, and as the razor edge cuts into it and it feels the pain, instead of going back into the hole it tries to escape the pain by going forward and so slits itself open all along the body. This is a dangerous practice however, for if the snake recovers it will become vicious and seek out humans.[45]

---

45. Personal communication.

7 Names referring to medicinal usage

Infusions from small pieces of bark (half the size of a thumb), pulverised in half a cupful of milk or broth, are used as purgatives for severe abdominal disorders . . . Finely ground dry bark is rubbed as an irritant into incisions on the skin for inflammation and pains in the chest. Ground bark, mixed with the dry root of an *Amaryllidaceae* species, is rubbed into incisions on dropsical swellings. Powder from the pulverised bark, mixed with bark of *unukani* (*Ocotea bullata*) and a little ginger, is blown through a small hollow reed into the womb to treat uterine disorders. Roots are used as purgatives and enemas for patients with fever . . . The fumes from ground leaves mixed with those of other Croton species on hot coals are inhaled for insomnia . . . Leaves are also used as steam baths for fevers and as deodorants. Unspecified parts are used for love charm emetics (Hutchings 1996: 166).

Introduction

In the opening quotation above, Hutchings gives some details about how parts of the tree Lavender Croton (*Croton gratissimus*)[1] are used by Zulu *izinyanga*. What is striking, perhaps, is how one plant can be used for so many different ailments and administered in so many different ways. In this chapter, we shall

---

1. With the Zulu names **umahlabakufeni**, **uhubeshane**, **inkubathi**, **isikhumampuphu** and **ilabele** (Boon 2010: 234).

look at Zulu names for plants that essentially reflect the same things referred to by Hutchings.

In Chapter 2 of this book we saw in Anna Pavord's narrative of the history of plant naming that it was only in the seventeenth century that the discipline of botany, with its specialists called botanists, came into being.[2] Prior to that almost all plant study was carried out by those involved in medicine and I quoted Pavord as saying:

> The apothecaries have a particular need to sort, name and categorise indigenous plants. Medicine is their business and plants the raw material from which they brew, distil and decoct their elixirs and tonics (2005: 5).

Even when we look at books relating to plant usage in South Africa in the second half of the twentieth century and today, we see this focus on the medicinal use of plants: for example the monumental 1932 and 1962 editions of *The Medicinal and Poisonous Plants of Southern and Eastern Africa* produced by Watt and Breyer-Brandwijk, Bryant's *Zulu Medicine and Medicine-Men* (1970), Roberts' *Indigenous Healing Plants* (1990) and Hutchings' *Zulu Medicinal Plants* (1996). Only comparatively recently, in 2000, was a book (Van Wyk and Gericke's *People's Plants*) about practical aspects of plant usage (food, drink, dyes, sources of craft material, etc.) published. To my knowledge nothing has yet been produced by way of a general book about all ethnobotanical usage in Zulu society and there have been no books at all (as far as I know) about the use of plants as protection from danger, and their use in esoteric areas of 'magic' and protective and love charms.

There is no question that the medicinal use of plants is an extremely important aspect in Zulu ethnobotanical usage. Zulu society esteems two professionals: the *inyanga* and the *isangoma*, both of whom are experts in the medicinal usage of plants, with the *isangoma* perhaps leaning more towards plant usage as psychological and spiritual intervention. Only the *inyanga yezulu* is a professional using plants for protection against natural dangers such as lightning and storms, and he certainly does not have the same status, profile

---

2. Pavord referred to the English 'plantsman' John Ray (1627–1705) whose six rules for a modern system of nomenclature "provided the vital underpinning of a new discipline which would later acquire a new name – taxonomy" (2005: 395).

and exposure as the *inyanga yokwelapha* (doctor of healing) and the *isangoma* when it comes to both popular and academic writing. And I am aware of no professional class at all in Zulu society featuring individuals who are highly trained experts in the use of plants for food or craft sources.

Problems in description and categorisation

In terms of both sub-divisions and sub-categories on the one hand and definition and explanation of terms on the other, we need to tread carefully. Ethonobotanical plant usage, and very often the Zulu names that refer to such usage, often fall between two potentially different categories, and can frequently be explained or defined in different ways. In many cases the use of a particular plant, or a medicine made from that plant, can be used for contrasting purposes and in widely different ways, and use can fit into a number of different conceptual frameworks. For example, an infestation of lice can be treated by the use of plant-based medicines that act as pesticides. Should the usage of these plants be discussed under 'medicinal usage' or 'practical usage'? Many sources claim in addition that lice are among a number of human afflictions actually directed against humans by witchcraft. So perhaps the usage of medicinal preparations against lice should be discussed under neither 'medicinal usage' nor 'practical usage', but under the general heading 'protective charms'. Then again, is there a conceptual difference between lice on the body that have occurred 'naturally' as opposed to an infection that has been caused by directed malice? Would one treat them differently? When one treats malice-directed lice, is one treating against lice or malice; against body vermin or witchcraft?

In the usage of medicinal plants described as follows, we shall often come across a statement such as 'may be used to cure or cause heart ailments'. For example, Hutchings says of the Natal Worm Bush (*Cadaba natalensis*) that "roots are used to make emetics and are also used to cure pleurisy, believed to be caused by witchcraft. They are also believed to be used in witchcraft to cause pleurisy" [my emphases] (1996: 111). Should these two diametrically opposed aims be treated as two sides of the same coin? Can one plant, or one medicinal concoction, indeed both cure and cause a particular ailment simultaneously? Do we not need two different catalysts for each goal – one an 'activator', the other a 'blocker' – and if so, should these essential elements not be factored into definitions and explanations?

Two sub-categories

This may be rather an artificial sub-division, but I have divided the names of plants that refer to medicinal use into two sub-categories:

- those that refer to the disease to be treated or to the afflicted body part – in other words, to the <u>target</u> of the medicine; and
- those that refer to the intended effect of the medicine – in other words to the <u>result</u> of the treatment.

We then need to add a third section (a sort of 'medical appendix'): the names of <u>medicines</u> rather than the names of medicinal plants.

Names referring to the disease or body part to be treated

Plant names under this heading all refer to plants that have some medical or curative use, and in each case the name refers to the particular disease to be cured, the body part affected, or the condition to be alleviated. For example, let us take the name **umfana-ozacile** (thin boy), the Tremble Tops (*Kohautia amatymbica*) that Pooley says is "used to improve appetite in infants" (1998: 202). In this case the name **umfana-ozacile** <u>refers</u> to a plant used to stimulate appetite, but the name is <u>based</u> on the actual condition of a child whose appetite needs to be stimulated because he is too thin.

A specific illness is named in **ikhambi-lesimungumungwane** (herb of measles), a name Pooley (1998: 208) gives for the scrambling shrub **Vernonia anisochaetoides*. Ndela Ntshangase says this plant is also used in the treatment of *uzagiga* (mumps).[3]

Another specific illness can be found in the name **isidala somkhuhlane** (dianthus of the cold/flu), a name for the Butterfly Lobelia (*Monopsis decipiens*) that Pooley tells us is used "to treat colds" (1998: 496). Ndela Ntshangase, in noting that the verb *dala* in the name also means 'create' or 'cause', explains that in order to <u>cure</u> a particular complaint, an important principle is to include a small proportion of whatever it is that <u>causes</u> that same complaint. Hutchings gives the Zulu name as a three-word phrase – **isidala ikhambi lomkuhlane** [*sic*] (i.e., Dianthus, a medicinal herb used for colds and flu). She notes that "whole plant infusions are taken in small doses for colds" (1996: 309).

---

3. Contributions from Ndela Ntshangase are all personal communications.

Other names refer to the body part that is affected. First, let us look at the gall-bladder: Pooley (1998: 318) gives the Zulu name **ikhambi-lenyongo** (herb of the gall-bladder) for the Wild Creeping Sunflower (*Aspilia natalensis*). Ndela Ntshangase points out that this plant is also known as **umphamephuce** (give it and take it away again), a reference to its use when the gall-bladder is excreting too much gall. The word *inyongo* (gall, gall-bladder) also occurs in the name **inyongwane**, which Hutchings records for *Corbichonia decumbens* (no English common name given), saying that "root infusions are taken for biliousness" (1996: 93). Another name based on the word *inyongo* is **ikhaphanyongo** (what drives away gall), given by Pooley (1998: 318) for the scrambling herb *Melanthera scandens*. She says root infusions are used as emetics.

From the gall-bladder to the bladder itself.[4] Hutchings says of the plant *Schizobasis intricata* that "bulb decoctions are taken for bladder pains associated with venereal disease" (1996: 29) and this is surely linked to the Zulu name **umthondo-wemfene** (baboon's penis) she gives for this plant.

Then we can consider the chest, in the name **umuthi wezifuba** (chest medicine). Pooley gives this as the name of *Ursinia tenuiloba* (no common name), "used to treat coughs" (1998: 332). Although Hutchings (1996: 332) does give this as a name for *Ursinia tenuiloba*, she says (correctly, in my opinion) that this is a general term for chest medicines rather than a name for a specific plant.

From chest to breasts: Hutchings (1996: 127) gives the name **umlungumabele** for the Spiny Splinter-Bean (*Adenopodia spicata*) and says that the roots are used for treating sharp pains in the breast. The element *amabele* (breasts) is easily discernible in the Zulu name, but the first element 'lungu' causes pause for thought. It is unlikely to be *umlungu* (white person) or *ilungu* (joint of finger), and it is quite possible that this is a mis-recording of the name 'umhlungumabele' where *amabele* is added to [*ubu*]*hlungu* (pain). However, neither 'umlungumabele' nor 'umhlungumabele' are recorded in Doke and Vilakazi.[5]

The symptoms of illness, rather than reference to specific body part, may form the basis of a Zulu plant name. Doke and Vilakazi provide us with the

---

4. If these are not the same thing?
5. Note that Zukulu et al. (2012) have previously been quoted as saying that the Small Knobwood tree (*Zanthoxylum capense*) was originally called 'umnugumabele' in Xhosa, but was renamed 'umlungumabele' because of the shape of a woman's breasts when wearing a European brassiere. I find this hard to believe.

name **udumesiswini** (rumble in the stomach) and say it refers to a "species of small creeper" (1958: 173). This plant is not mentioned in Pooley or Hutchings. Ndela Ntshangase notes that this plant is used as a purgative and makes much abdominal noise.

From abdominal noise we move easily to flatulence. Hutchings gives the name **umoyawezwe** (wind of the nation) for the Blue Cress (*Heliophila subulata*) and says "roots are sometimes chewed by healers and spat upon the stomach of patients whose stomachs are distended by wind" (1996: 108).

Quite a different symptom is found in the plant name **ilozane** (restlessness in sleep). Pooley gives this name for the Creeping Tephrosia (*Tephrosia macropoda*) and says "used to treat . . . fevers" (1998: 392).

A far more general symptom of illness is seen in the name **inzwabuhlungu** (feel the pain). Pooley gives this name for the Beautiful Senecio (*Senecio speciosus*), saying it is "used to treat chest complaints, dropsy, headaches, & for poultices" (1998: 444). This Zulu name surely has commercial value, especially tailored for advertising jingles: 'when you feel the pain, take *Feel-the-Pain*'.[6]

We look at poisoning when we come to the Hairy Caterpillar-Pod (*Ormocarpum trichocarpum*). It has as one of its Zulu names **umsindandlovana**, derived from *sinda* (recover) + *indlovu* (elephant) and the diminutive suffix *-ana*. Boon says of this plant, "'Even an elephant will recover' because bark used as an emetic to treat poisoning: *Mundulea sericea* with similar bark has the same Zulu name for this reason" (2010: 162). The other name he gives for *Ormocarpum trichocarpum* is **isithibane**, derived from the Zulu verb *thiba* (ward off), and this clearly also refers to its use.

Let us note now a name referring to sores and wounds. Hutchings records the name **ikhambi-lesilonda** (herb of the wound, sore) for both *Cheilanthes viridis* and the closely related *Pellaea rufa* (no English common names given) and says "used as a household remedy . . . the Zulu name indicates that the plant is used for sores and other skin complaints" (1996: 8, 9).

Body part <u>and</u> symptom are referred to in the two names of the Quilted-Leaved Vernonia (*Vernonia hirsuta). Pooley gives the names **ikhambi-lenyongo** (herb of the gall-bladder) and **uhlunguhlungwana** (variety of little aches and pains). This plant is used "to treat colic, sore throats, coughs, headaches, and rashes" (1998: 498).

---

6. It would sound better in Zulu.

There are a number of plant names with an underlying meaning that refers to their treatment of hysteria and madness and it seems sensible to assemble them under their own sub-heading:

Hysteria and madness

The wild yam (*Dioscorea dregeana*), "used in traditional medicine to treat insanity, fits" (Pooley, 1998: 514), has three Zulu names that refer to the conditions treated by a medicine based on ingredients from this plant: **ilabatheka**, **undiyaza** and **udakwa**. Let us look at each of these in turn:

- **ilabatheka**: Doke and Vilakazi say that **ilabatheka** is used as the name for "two species of *Liliflorae* with poisonous qualities", these being *Dioscorea dregeana* and **Hypoxis latifolia*, but the basic meaning of the word is "[any] medicine for causing madness or excitement" (1958: 444). The noun is derived from the verb *labatheka* (be carried away with longing, feel a strong desire for something). Hutchings (1996: 55) gives *Hypoxis latifolia* as a synonym for *Hypoxis colchicifolia* and records the Zulu names **ilabatheka**, **ilabatheka-elimnyama** (black labatheka) and **igudu**. Among other medical usages, infusions from this plant are administered for hysterical fits. The association of hysterical fits with drunken behaviour takes on an extra dimension with the name **igudu** (< *igudu* (horn used for smoking wild-hemp));[7]
- **undiyaza**: as Doke and Vilakazi note, the verb *ndiyaza* means 'be stunned, confused, giddy'. From this is derived the plant name **undiyaza** "species of forest climber, *Dioscorea dregeana*, which causes madness" (1958: 539). The verb itself is derived from the ideophone *ndiya* (of being stunned, confused, giddy) with the verb-forming suffix *-za*. The same ideophone is reduplicated as the noun stem *isindiyandiya*, which in addition to meaning 'dizziness, giddiness, perplexity, confusion', is also used as the plant name **isindiyandiya**, referring to the tree *Bersama lucens*, "taken as a charm to confuse one's opponents in court" (539); and
- **udakwa** (being drunk): the perceived drunken nature of hysterical fits and/or insanity is reflected in three different forms of the verb stem *-dakwa*. As we have just seen, Pooley gives the form **udakwa** for *Dioscorea dregeana*. Hutchings gives **isidakwa** for *Dioscorea dregeana*

---

7. Cannabis, marijuana, dagga.

and assigns **udakwa** to *Dioscorea diversifolia*, both apparently known by the same English common name of Wild Yam. The medicinal usage of both species, according to Hutchings, is almost exactly the same, with *D. diversifolia* recorded as "tubers are used to treat hysterical fits" and *D. dregeana* as "tubers are used for hysterical fits and to cure insanity" (1996: 57). Doke and Vilakazi offer the third form of the name in **uliyadakwa** (it is becoming drunk) for *Dioscorea diversifolia*, referring to the plant as a "species of forest climber" (1958: 460).

Boon gives the Zulu name **umgogo-wezinhlanya** as one of five names for the Dwaba-Berry (*Monanthotaxis caffra*), saying that it is "traditionally used against hysteria and bad dreams" (2010: 102). Hutchings (1996: 104) and Doke and Vilakazi (1958: 253) make it clear that the name should be **umgogi-wezinhlanya**. From the verb *goga* (obstruct, prevent, disable) the noun *umgogi* is derived, with the meaning 'herbalist's charm medicine', and from this is derived **umgogi-wezinhlanya** (madmen's charm). Doke and Vilakazi say "herbalist's name for the *Popowia caffra*, species of tree used as emetic against dreaming" and Hutchings, who gives *Popowia caffra* as a synonym for *Monanthotaxis caffra*, says "roots are smoked for hysteria and bad dreams".

Two more plant names from Doke and Vilakazi enlarge on the madness metaphor. Also based on *uhlanya* is **uqumbahlanyana** (< *qumba* (be sulky, angry) + *uhlanya* (madman) + dim. *-ana*) (1958: 715). Doke and Vilakazi give this as a "species of flowering composite, *Senecio lanceus*" (1958: 764). Hutchings (1996: 329), who gives the name as 'qumbahlanga', makes no mention of its use against hysteria. Doke and Vilakazi also give us the name **abaphaphi**, derived from the verb *phapha* (be uncontrolled, mad, wild). Their primary meaning for this name is "spirits of madness"; their secondary meaning "herbalist's name for *Euphorbia ingens*" (1958: 648). Again, there is no mention of treating hysteria in Hutchings (1996: 175), who gives this name as 'umphapha'. The closest she gets is mentioning that the Sotho use this plant to treat dipsomania.

Boon gives the two Zulu names **uqhume** and **isiphahluka** for the False Horsewood (*Hippobromus pauciflorus*) and says "bark, leaves and roots [are] widely used . . . for hysteria" (2010: 326). Doke and Vilakazi see the name **u(lu)qhume**, "species of small bush, *Hippobromus alatus*" (1958: 704), as derived from the verb *qhuma* (burst out, explode) and the relevance of the name **isiphahluka** is clear in its derivation from the verb *phahluka* (blurt out, speak without thought). Hutchings (1996: 190), who gives *Hippobromus alata* as an earlier synonym for *Hippobromus pauciflorus*, also recognises *uqhume*, gives *isiphahluka* as 'isiphaluka' and adds the name **umfazi-othethayo** (scolding, nagging woman). She notes the use of this plant for hysterical fits and quotes healers from the Valley Trust as saying that a plant called 'uqhume' is used in the treatment of psychiatric disturbances.

In the same line is **ingulamlomo** (what opens the mouth), an unidentified medicinal plant that Doke and Vilakazi say (1958: 564) is used against hysteria. They add that it is also used as a charm to aid in law cases.

Shouting out aloud need not be the result of an attack of hysteria. Some people apparently shout in their sleep and for this the obvious cure is the plant **ummemezi** (the shouter). Doke and Vilakazi say this is a "species of plant whose bark is sold by Native herbalists to cure shouting in sleep" (1958: 498). Boon (2010: 446) identifies the plant as the Fluted Milkwood (*Chrysophyllum viridifolium*).

A person having a hysterical fit can also be seen as behaving in a foolish or idiotic manner. This is reflected in the plant name **umluthu**, for which Doke and Vilakazi (1958: 471) give the primary meaning 'fool, idiot, imbecile' and a secondary meaning 'herbalist's remedy for hysterical fits'. Hutchings (1996: 264) gives this name for both *Vitex rehmannii* and **Vitex wilmsii*, and indicates that both species are used in the treatment of hysterical fits.

Doke and Vilakazi give the name **umahayiza** (< *hayiza* (rave, be hysterical)) saying this is a "species of orchid, prob. *Lissochilus*, used as an emetic in treating hysteria" (1958: 477).

A final entry under the sub-heading of hysteria reminds me of the crowds of hysterical girls who used to be seen on television in the 1960s and 1970s waiting for their favourite pop idols to arrive at airports. The name, given by Doke and Vilakazi is **usokalakwazulu** (lover-boy from KwaZulu) and they say it refers to a "medicinal compound believed to cause hysteria in young girls" (1958: 763).

Biting creatures and pests

When lice are a problem, one needs a preparation made from **ikhambi-lentwala** (herb of lice), a name that refers to the Trailing Daisy (*Microglossa mespilifolia*). Hutchings also records this name for the epiphytic fern *Microgramma lycopodioides* and says:

> Whole plants are used against pubic lice on humans and also to prevent them from being transferred from one person to another . . . Such lice are traditionally reputed to be inflicted through sorcery (1996: 9).

Ndela Ntshangase confirms this when he says that plants with the name **ikhambi-lentwala** are used both to <u>control</u> and <u>cause</u> lice.

Snakes and the danger of snakebite have always been problems in southern Africa, and it should be no surprise to find plant names referring to these. In Chapter 8 we shall look at names that refer to plants used as charms to keep snakes away. Here we look at the problems caused when snakes have <u>not</u> been successfully kept away.

Pooley gives the Swazi name *litinyolemamba* (tooth of the mamba) for the perennial climber Traveller's Joy (*Clematis brachiata*),[8] which is used "as a snakebite remedy" (1998: 140). When snakebite causes a burning sensation, it would be useful to treat the bite with **umakuphole** (let it cool down). Doke and Vilakazi give this as a "species of common meadow herb . . . *Pentanisia variabilis* . . . root used for snake-bite" (1958: 479). The poison injected by certain species of snakes is singled out for special reference in **ikhambilesihlungu** (herb of poison). Pooley gives this as a name for the Fig-Leaved Ipomoea (*Ipomoea ficifolia*), used "to treat snake-bite" (1998: 420). And then again, the pain caused by snakebite is part of the underlying meaning in **ubuhlungu-bemamba** (pain of the mamba [bite]), one of Pooley's Zulu names for *Clivia miniata* (St John's Lily), which is used "to treat . . . snakebite" (1998: 38).

The mamba appears again in **umshayimamba** (hit the mamba), one of several names given by Hutchings (1996: 248) for the Monkey Ropes (*Strophanthus speciosus*). Hutchings tells us that for snakebite, powdered roasted roots are used, and sometime roasted snake heads are included in the potion.

Our last 'snakebite-name' is **umanqanda** (that which turns away, checks, prevents). Pooley gives this name for the Humped Turret-Flower (*Asclepias gibba*) "used traditionally to treat snakebite" (1998: 170). Hutchings says of the medicinal use of this plant that:

> Dried powdered roots are licked off the back of the hand for snakebite, causing the patient to vomit and bring up the foam believed to be caused by the poison of the snake . . . It was traditionally believed that snakebite caused a froth to form in the throat which caused death when it reached the heart (1996: 253).

---

8. The English name of this plant, Traveller's Joy, is rather ironic in the light of the Swazi name!

Ndela Ntshangase has some interesting comments to make about snakebite remedies. First, he notes that although the phrase *isihlungu senyoka* literally means 'snake poison', it can also be used to refer to snakebite remedies. Second, he notes that another Zulu name for the Clivia mentioned above is **umayime** (let it stop).[9] The significance of this name is a double one: when you take *umayime/ubuhlungu-bemamba*, which you *khotha* (lick from the hand), as the snake poison moves up the body from the site of the bite, the remedy moves down until it meets the poison and both poison and remedy stop right there. Furthermore, once you have taken this 'muthi', forever afterwards whenever your shadow falls on a wild snake it will freeze right there and you can safely take a stick and kill it. Incidentally, says Ntshangase, when taking any plant-based 'muthi' for snakebite it is always a good idea to grind up a little snake bone in it.

It is not clear whether the symptom or the effect of the medicine is referred to in **umthundangazi** (what urinates blood),[10] the Blue Sweet-Berry (*Bridelia cathartica*), but clearly it is the end result that is referred to in **umphunziso** (what causes a miscarriage in animals), a name for the Forest Bush-Cherry (*Maerua racemulosa*).[11] The name is derived from the verb *phunza* (have a miscarriage, abort, used of animals only). Ntshangase notes that this tree is something only an *umthakathi* would use, so one should avoid using it in any way otherwise one will be suspected of witchcraft.

Again, is a 'big heart' a problem that needs to be cured, or is it a desirable end result in **inhliziyonkulu** (big heart)? This is the Zulu name Pooley gives for the Common Wild Pear (*Dombeya rotundifolia*) "used medicinally for heart problems" (1993: 314). Hutchings also says "inner bark used for heart problem" (1996: 200) and Doke and Vilakazi (1958: 569), who give the name as **unhliziyonkulu**, say the inner bark is used medicinally for faintness, so perhaps the answer is that this plant is used for curing a medical problem rather than intending to imbue the user with 'big-heartedness'.

From urinating blood and problems with the heart, we come to blood itself. Derived from the noun *ingazi* (blood) are three Zulu plant names: **ungazi**, **ungazinde** and **ungazini**. Their primary and secondary meanings, as

9. Pooley also gives this name.
10. Although Ntshangase suggests that this refers to urine tinted with the red-colouring in the "reddish-purple to black" fruits (Boon, 2010: 230).
11. Does this tree in fact contain an abortifacient? Hutchings does not mention the tree at all.

given by Doke and Vilakazi (1958: 550) show links between plant, medical condition and intended effect when used in witchcraft. The primary meaning of **ungazi** is "species of forest tree whose bark exudes a red sap, said to cause purpura[12] and fatal haemorrhage when used by witches"; and the secondary meaning is simply "fatal haemorrhage". The primary and secondary meanings of **ungazinde** are the other way around, with the primary meaning "fatal haemorrhage or other disease caused by bleeding" and the secondary meaning "species of forest tree whose bark exudes a red sap". There is only one meaning given for **ungazini**, namely "species of forest tree whose bark exudes a red sap". Sadly, plant names given in Doke and Vilakazi with no clear indication of a definite botanical species seldom find their way into Hutchings (or indeed, Pooley, Boon, Moll and the others) and so we are unable to check whether this red sap-exuding forest tree actually does have chemical properties that affect the blood, or whether the link to fatal haemorrhages is simply a result of the blood-like appearance of the red sap.

Both problem <u>and</u> cure can be read into **umthobo** (scrofulous body swelling, herb used for poultices), a name for the Weeping Wattle (*Peltophorum africanum*).

A very 'general purpose' name is **umlulama** (what recovers from illness), used for both the Butterspoon Tree (*Cunonia capensis*) and the Wild Honeysuckle Tree (*Turraea floribunda*), "used medicinally for rheumatism and heart ailments" (Pooley 1993: 200). Hutchings records this name for the plants *Deinbollia oblongifolia* (1996: 189), *Maytenus acuminata* (1996: 182), **Myrica serrata* (1996: 73) and *Nuxia floribunda* (1996: 240). Hutchings notes the following illnesses and conditions from which a patient can recover when using **umlulama**:

- dysentery and diarrhoea, strengthening after death of kraal member, headaches (*Deinbollia oblongifolia*);
- stomach ailments (*Maytenus acuminata*);
- cold, coughs, headaches and dysmenorrhoea (**Myrica serrata*); and
- fevers, coughs, indigestion, influenza and infantile convulsions (*Nuxia floribunda*).

---

12. *Collins Dictionary of the English Language* explains this as "any of several blood diseases causing purplish spots or patches due to subcutaneous bleeding" (1986: 1241). The word is derived from the Greek *porphura*, a shellfish yielding a purple dye.

The recovery implicit in the name **umlulama** is more spiritual and psycho-somatic when Pooley's *Cunonia capensis* and *Turraea floribunda* are used medicinally. Hutchings records both **umlulama** and **umlulama-omkhulu** (big umlulama) for *Cunonia capensis* and says "strengthening medicine made from the plant is taken after the death of a kraal member" (1996: 115). The significance of her additional name for this plant, **umuthi wokuzila** 'mourning medicine), is obvious. *Turraea floribunda*, says Hutchings, is also used to strengthen people after the death of a kraal member and to prevent fearful dreams, while "roots are used by diviners to enter the neurotic state needed for divining dances" (1996: 156). It is also used for more purely medical reasons, to treat rheumatism, dropsy, heart disease, and swollen and painful joints. She gives the names **umlulama, umlulama-omncane** (little umlulama) and the curious variant 'ululame'.

Intended effect of treatment

The plant names under this heading all make reference to an intended effect when ingredients from the plant are used to make medicines.

Impotence and infertility

One of the Zulu names for the Flame Lily (*Gloriosa superba*), preparations of which are intended to cause erections in males suffering from impotence, is **isimiselo** (cause to stand, remain firm). Hutchings mentions a number of related uses:

> Powdered corms are sometimes taken in whey for men for impotency or mixed with food to be eaten by husband and wife for barrenness or to ensure the desired gender of a child . . . Corms are selected from their resemblance to male or female genitalia . . . They are also used as aphrodisiacs and are sometime taken as an emetic charm by a young man to cause an indifferent girl to appear pregnant until she returns his affection (1996: 26).

Similar aims are found in the name **ubangalala** (cause to sleep/have sex): Doke and Vilakazi state that this refers to various species of *Rhynchosia* that cause sexual excitement. Hutchings, giving the name **ubangalala** for various *Rhynchosia* species, says "roots of species known as *ubangalala* are cooked in milk and taken in mouthfuls as aphrodisiacs" (1996: 145).

The Scarlet Eriosema (*Eriosema distinctum*) is known in Zulu as **ubangalala olukhulu** (large bangalala). Hutchings says that decoctions from this plant "are used for urinary ailments and for impotence" (1996: 146). The Narrow-Leaved Eriosema (*Eriosema salignum*), which, says Pooley, is "used to treat impotence . . . and to stimulate bulls in spring" (1998: 62) also uses the name **ubangalala** as well as **uphondongozi** (horn of danger), an extremely suggestive name. Hutchings gives both **ubangalala** and **ubangalala-oluncane** (small bangalala) for this plant and says "hot milk infusions of roots or cold water infusions of rootbark are taken in small doses at night and in the morning for impotence" (1996: 146).

The bull is associated with male strength, prowess and potency in many cultures, no doubt because of the renowned length of that legendary organ, the 'bull's pizzle'. We see this association in the Zulu name **umhlabankunzi** (prick the bull), a name given to the Heart-Leaved Eriosema (*Eriosema cordatum*), which is used "to treat impotence and to stimulate bulls in spring" (Pooley 1998: 274).

Also used for impotency (among many other ailments) is the plant *Capparis tomentosa*, the Woolly Caper Bush, and again we find the bull metaphor. Hutchings gives the name **inkunzi-ebomvu** (red bull) as one of five Zulu names. We see the same reference in **inkunzi** (bull), one of five Zulu names Hutchings gives for the Lavender Tree or Lemon Verbena (*Heteropyxis natalensis*), of which she says "bark used to treat impotence and as an aphrodisiac" (1996: 219).

The bull appears yet again in **umvusankunzi** (what rouses the bull), the Forest Num-Num (*Carissa bispinosa*). Although, according to Hutchings, unspecified parts are reputed to have aphrodisiac properties, it is likely that the underlying meaning of this name is less metaphoric than the 'bull-names' above: Hutchings reports that this plant is

> reported to be used in traditional ceremonies by Swazi paramount chiefs to make sacrificial bulls fierce . . . [t]he bulls are struck by a switch so that they will not be easy victims for the warriors, who have to overcome the animals with their bare hands (1996: 244).

A final aside on the word *inkunzi* is its use in **inkunzithwalitshe**, one of several names Hutchings gives for *Curtisia dentata*. This compound noun has three elements: *inkunzi* (bull) + *thwala* (carry) + *itshe* (stone); i.e., 'the bull carrying a stone'. Hutchings mentions that the plant is used as an aphrodisiac, but even so the underlying meaning seems obscure.

Although there appears to be no reference to impotence in the underlying meaning of the name **intwalabombo** (that which climbs the ridge), a name for the *Rubia cordifolia* plant,[13] Berglund's mention of this plant is not without interest:

> Men who are unable to attain erection are given infusions of *inthwalabombo* [*sic*] (*Rubia cordifolia*, a species of climbing plant said to be soft and bendable without breaking off, associated with the male member). 'The medicine enters the body. It goes to that place (male member). It does its work there. The man goes to his wife and finds that he can work nicely' (1976: 354).

Hutchings says of this same plant (*Rubia cordifolia*) that "root decoctions are taken at bedtime for the treatment of impotence" (1996: 301). She gives the Zulu name **intwalalubombo** as well as the name **impindisa**. This name, according to Doke and Vilakazi is derived from the verb *phindisa* (cause to repeat) and is used for a "species of climbing plant, *Rubia cordifolia*, used against impotence" (1958: 663).

Medicines intended to counter impotence in men have their equivalents in those intended to relieve women suffering from infertility, as reflected in a number of names based on the noun *ihlamvu* (seed, pip, kernel). The first is **uhlamvulwentombazane** (< *ihlamvu* (seed/pip/fruit) + *lwentombazane* (of the girl)), the Littonia (*Littonia modesta) that Pooley tells us is used "for infertile women and if a female child is desired" (1998: 26). Curiously enough, when the same word *ihlamvu* is used as a name without 'lwentombazane', the aim is to have male children: Hutchings says the name **ihlamvu** is used for various *Habenaria* species and "decoctions from cut up whole plants are kept in clay pots and small quantities are taken every morning by couples who desire sons and only have daughters" (1996: 65). The same name, **ihlamvu**, is given by Hutchings for the plant *Disa aconitoides*, of which she says "root infusions

---

13. Sticky-Leaved Rubia to Pooley (1998: 558), but Dyer's Madder to Doke and Vilakazi (1958: 811). 'Madder' is a word one seldom comes across nowadays. *Collins Dictionary of the English Language* (1986: 923) says it is derived from the Old English *maedere* and that the word is used for (1) any of several rubiaceous plants of the Genus *Rubia* (2) a dark reddish-purple dye obtained from fermentation of the roots of this genus and (3) a 'red lake' obtained from alizarin, used as a pigment in inks and paints. I recall coming across the pigment 'rose madder' in school water colour paintboxes in the early 1960s, but *Collins* does not record the term.

are taken by women as emetics to promote conception" (1996: 66).[14] And then again, Hutchings refers to **uhlamvu-lwabafazi** (seed of the women), a name for the plant *Eulophia cucullata* of which we note "root infusions [are] taken by . . . married couples when the wife is feared barren" (1996: 69).

It is not clear whether it is men, women or both who need to take medicines derived from the Ox-Eye Daisy (*Callilepis laureola*), used "to treat . . . infertility" (Pooley 1998: 216). The Zulu name **impila** (life, living) is derived from the verb *phila* (be alive, be healthy).

We have looked above at a number of Zulu plant names referring to their use in combating impotence and sterility. If medicines based on these plants are found to be successful, then the happy husband and wife may need to look for two more plants, one named **umsekelo** and the other **isiphuthumane**.

Pooley (1998: 280) gives **umsekelo** as a Zulu name for the *blouboktoutjie* (*Pyrenacantha scandens*) and Doke and Vilakazi, who explain the name as derived from the verb *sekela* (prop up, support), say it is a "medicine to prevent miscarriage" (1958: 725). Hutchings confirms that "cold water root infusions are taken during pregnancy to prevent miscarriages" (1996: 187) and adds that such root infusions may also be used to ensure an easy birth, which leads us to the next name. **Isiphuthumane** (hasten) is a name Pooley gives for *Aloe cooperi*, "used in traditional medicine for easy birth" [i.e., to hasten the birth] (1998: 32). Hutchings, who gives the name as **isiphuthumana**, cautions that the roots "are used from the eighth month only as earlier use in pregnancy may cause abortions or miscarriage" (1996: 33).

The medicines we have discussed above are designed to increase potency and fertility. The opposite effect is intended in medicines given to one's enemies and these may be based on the Sensitive Plant (*Mimosa pigra*) with its two names **imbuna** (wither, shrivel, droop) and **umazifisa** (self-desire)[15] "used as an ingredient in preparations designed to make an enemy impotent" (Pooley 1998: 388). The use of this plant in concoctions designed to render enemies impotent is based on 'sympathetic magic': the Sensitive Plant, as we have noted earlier in this book, has leaves that collapse or droop when touched, and this is the effect desired for the enemy's penis.

The verb *lala*, found in **ubangalala** above with the meaning 'have intercourse', is used with its more common meaning 'to sleep' in **umalala**

---

14. Pooley (1998: 364) also gives the Zulu name **umashushu** (strong desires) for this plant.
15. Also given by Doke and Vilakazi (1958: 489, 518) as **umzifisi**(o-).

(sleeping), the scrambling shrub *Osyridicarpos schimperianus*. Hutchings says it is "used to stop fighting and to make babies sleep" (1996: 83).

Cooling medicines

All the medicines described above to combat impotency and fertility have the effect, in Zulu thought patterns, of creating heat. There are a number of medicines with the opposite aim, namely to cool things down and this aim is frequently referred to in the name of the plant.

Pooley gives both **makuphole** (let it cool down) and **icishamlilo** (put out the fire) for the Broad-Leaved Pentanisia (*Pentanisia prunelloides*) that is "used to treat a wide range of ailments from stomach pains to haemorrhoids" (1998: 492). Hutchings, who gives the names **icishamlilo** and **icimamlilo** (both of which mean the same thing) for this plant, details the cooling use:

> Pounded roots are applied to burns and used in poultices for inflammation and swollen joints . . . Leaf poultices or hot root decoctions are applied to painful swellings, rheumatic parts, sprains, sores and for fevers (1996: 298).

Also from the verb *phola* (cool down, be cool) is the name **umaphola** (being cool) given by Pooley for *Berkheya speciosa* (no English common name), which is "used to bathe sore eyes" (1998: 336).

The verb *phola* is used in its causative form *phozisa* (make to cool down) in the name **umaphozisa** given by Pooley for *Senecio serratuloides*, "used to treat burns" (1998: 328). Pooley and Hutchings (1996: 330) both give the unusual English common name Two-Day Cure for this plant, clearly an adaptation of the Zulu name **insukumbili** (two days). Hutchings adds to the cooling effects of this plant, detailing its use as a drawing agent for purulent sores as well as the use of powder from charred roots applied to burns. Smith (1966) surprisingly does not mention Two-Day Cure in his list of the common names of South African plants.

Continuing our examination of 'cooling names', we note **umancibilika** (melt, cool down), a name Pooley gives for the Spotted Knotweed (*Persicaria lapathifolia*), which is used "to treat venereal disease" (1998: 376).

An interesting name under this heading is **umlwandle** (< *ulwandle* (sea, ocean)). Doke and Vilakazi give this as a name for a "species of herb, *Hermstaedtia* [*sic*] *elegans*" (1953: 471). Hutchings identifies *Hermbstaedtia* *elegans* as an earlier synonym for *Hermbstaedtia odorata*, with the Afrikaans names *katstert* or *rooikatstert*. Pooley (1998: 50) makes it clear that this is not a littoral plant, so *umlwandle* is not a 'habitat name'. I conclude, then, despite there being no such indications in the literature, that *umlwandle* refers to the cooling powers of the ocean. It has been noted previously, within the context of the symbolism found in many *imithi*, that because no matter how hot the day ocean water is always cool, medicines made with either sea water or sea creatures, or both, can be used to cool down medical conditions characterised by heat. Although Hutchings only gives the Zulu medicinal use as "roots used . . . as a cleansing stomach wash" (1996: 88), she does mention that in Botswana root infusions are used for gonorrhea, so like **umancibilika** above, this plant is used to soothe the burning caused by venereal diseases.

Similar intentions, to 'calm down', 'cause to subside' and so on, are found in the names **idambisa** and **idambiso**, both derived from the verb *dambisa* (cause to subside). Pooley gives **idambisa** as the name of the Common Kalanchoe (*Kalanchoe rotundifolia*), which is "used in traditional medicine as an emetic" (1998: 54); and **idambiso** for *Cyphostemma natalitium* (no common name), "used for poultices" (1998: 282).[16]

A very similar underlying meaning is found in the name **ibohlololo** (< the ideophone *bóhlololo* (of calm after storm, of dying down of fire or anger)), which Pooley gives for the Aptenia (*Aptenia cordifolia*) that is "used as poultice and anti-inflammatory" (1998: 378). Doke and Vilakazi (1953: 81) give this as the name of three species of plants (**Aptenia cordifolia, Achyropsis avicularis* and *Senecio speciosus*) and Hutchings (1996: 404) has the same. For each plant, use as poultices is mentioned.

Treating dysentery and stomach disorders

A fairly straightforward link between medical treatment and the underlying meaning of a name can be seen in **umshekisane**. Doke and Vilakazi state that this name is derived from the verb *sheka* (pass loose stools as in diarrhoea > *shekisa* (cause to pass loose stools)) and give it as the Zulu name for a "species of tree, *Euclea undulata*, with small edible berries, roots used as a powerful purgative enema" (1958: 736). Hutchings gives **umshekisane** as the

---

16. Doke and Vilakazi say of **idambiso**: "species of wild vine used for poulticing, *Cissus connivens, C. lanigera*" (1958: 138).

name of *Euclea undulata*, the <u>female</u> plant of *Euclea crispa*,[17] the plant *Euclea divinorum* and the plant *Euclea natalensis*. Among many other medicinal uses, Hutchings records **umshekisane** as being used in the following ways and the underlying meaning is clear:

- "bark and leaves are used as purgatives . . . and bark infusions are administered as enemas" (*E. undulata*);
- "as an enema for stomach disorders" *(E. crispa)*;
- "fruit is taken as a strong purgative" (*E. divinorum*); and
- "preparations are made from a . . . length of rootbark, infused or slightly simmered in a couple of milk tins of warm water are taken or, mixed with more warm water, used as an enema for stomach disorders" (*E. natalensis*) (1996: 232).

Somewhat drastic treatments for dysentery are suggested in the underlying meaning of **uvimbukhalo** (what blocks up the loins), a name for the Leafy-Flowered Ipomoea (*Ipomoea crassipes*), and **umsongelo** (< *songela* (harden, become constipated)), a name for the Stalk-Flowered Pelargonium (*Pelargonium luridum*), both used to treat dysentery.

Another treatment for dysentery suggested by the Zulu name may not make medical sense to everybody: **isikhubabende** (what interferes with blood clots), *Indigofera eriocarpa* with no common name (Pooley 1998: 56). Doke and Vilakazi, responsible for the translation 'what interferes with blood clots', say this is "several species of dwarfy [*sic*] plants (*Indigofera* spp.) having woody carrot-shaped roots, used for dysentery" (1958: 407). Hutchings gives **isikhubabende** as a name for four *Indigofera* species: *I. eriocarpa, I. hilaris, I. micrantha* and *I. velutina*. Of all four she says, "the above species are all reported . . . to be used in remedies for dysentery" (1996: 136).

For Berglund the plant **isikhubabende**[18] has quite different symbolic values and medicinal uses. The root of this plant, he says, is red like that of **intolwane** (**Elephantorrhiza burchellii*)[19] and "red, the colour of blood and frequently related to menstruation and pregnancy, as a symbol, plays an important role in

---

17. The male plant being **idungamuzi**.
18. Given by Berglund as 'isikhubadende', clearly an error in transcription from written notes.
19. *Elephantorrhiza elephantina*: this plant is often found in the praises of a red bull, *uJamludi obomvu njengentolwane* (Jamludi who is red like the *intolwane* shrub) (Pooley 1998: 150).

fertility and pregnancy". So rather than being used to treat dysentery, the plant is given to expectant women at regular intervals during the time of pregnancy to "make the child strong" (1976: 356).

Miscellaneous medicinal aims

Doke and Vilakazi give the name **umabelebuce** (shriveled breasts) and say this refers to a "species of leguminous herbs whose roots are used medicinally by a woman after losing a baby she has been nursing" (1958: 473). Hutchings identifies **umabelebuce** as *Lotononis corymbosa* and says "roots are used for women who have lost suckling babies" (1996: 133). Hutchings also gives the name **umamatheka** (smile) for this plant and records "used as love charms".

Pooley gives the name **umnandi** (tasty) for *Syncolostemon parviflorus*.[20] This is "used in traditional medicine to treat loss of appetite" (1998: 188) (cf. **umfana-ozacile** (thin boy) above), so presumably the plant is used to make food attractively tasty in some way for those with no appetite, or there is a metaphorical process going on in the name that is not quite clear. Hutchings adds no clarity with "Milk infusions are administered as enemas to children and as emetics to adults for loss of appetite" (1996: 272).

Pooley gives the name **uguqukile** (having turned) for the Dwarf Hibiscus (*Hibiscus pusillus*), which is used "to treat bad dreams" (1998: 286). Presumably medicine from this plant will turn the bad dreams into good ones, or at least turn them away.

The last name in this sub-section is **udoloqina** (knees get strong), a name given by Pooley (1998: 438) for the Wild Scabiosa (*Scabiosa columbaria*) "taken as a tonic at the commencement of each new season" (Doke and Vilakazi 1958: 166). Hutchings (1996: 302) mentions neither this name nor this usage in her entry for *Scabiosa columbaria*.

Names of medicines

We end this chapter with a random sampling of names for unidentified medicinal plants, names that Doke and Vilakazi stress are 'herbalists' names', and the names of medicines, remedies and charms that are not necessarily the names

---

20. The English common name she gives for this plant is the confusing Small-Flowered White or Pink Plume.

of plants themselves. Most of these names do not appear in Pooley, Hutchings, Boon or other works of a botanical nature because being either unidentified species, or the names of entities that are not plants, they cannot be assigned to specific plant species. The question may be asked, then, why they are being included in this chapter, or indeed in a book on Zulu plant names, if many of them are not the names of plants. There are two answers to this question: one is that it is frequently difficult to distinguish between the use of a Zulu word to refer to a plant and the use of exactly the same word to refer to a medicine made entirely or primarily from that plant. The other answer is that although it is difficult, if not impossible, to pin down plant species in the following names, they are all names that are interesting, even intriguing, in their own right. When a name for an unidentified species of plant used as a love charm has the underlying meaning of 'Piece of firewood, I am thirsty', I for one find it impossible to leave it out of a book on Zulu names of plants.

There has been no attempt to categorise the following names and in most cases it is difficult to make even a pertinent comment. The names must speak for themselves. The numbers indicate a page number in Doke and Vilakazi (1958):

- **umalalangenxele** (what lies on its left side): species of medicinal plant (479);
- **umhlalamthini** (what sits in the tree): type of love charm (315);
- **uhlabazihlangana** (what stabs on meeting): herbalist's love charm;[21] cf. **uhlabe-hlangana**: "sp. of *Rhamphicarpa*: a hemiparasite in the grassveld, used as a medicinal charm" (310);
- **ugushamoya** (what is carried with the wind): type of herbalist's remedy; cf. **ugushanomfula** (what is carried away with the river) a type of herbalist's remedy (279);
- **ugobabahleke** (that which bends people that they laugh): type of herbal medicine (251);
- **ingobamakhosi** (tamer of chiefs): 1) regiment formed by Cetshwayo; 2) type of herbal love charm (251);
- **umgodweni** (in the hole): a herbalist's emetic (253);
- **umgogi-wezinhlanya** (madmen's charm): herbalist's name for the *Popowia caffra*, a sp. of tree used as an emetic against ('against'?) dreaming (253);

---

21. Shades of Cupid's arrow here!

- **ukhalamasoka** (what humiliates the young man): sp. of plant, whose roots are used as a love charm (376);
- **umgqunkqiso**: medicinal charm used to cause loss of attractiveness < *qunkqa* (fade, become discoloured) (268);
- **inkiphadliso** (< *khipha* + *idliso* (poison) = what takes out poison): a herbalist's remedy for chest trouble (394);
- **u(lu)khuningomile** (piece of firewood, I am thirsty): sp. of plant used as a love charm (466);
- **undinabembuka, umdinabemkhangala** (who annoys them while they look at him): baboon fat used as a charm (538);
- **umanyokana** (little snake): sp. of medicinal herb used for stomach troubles (485);
- **ungezakanye**: herbal love charm (< *ingeza* (purifying medicine, love charm to make one attractive) < *geza* (wash) + *kanye* (at one time)) (555);
- **abaphaphi** (< *phapha* (be uncontrolled, mad, wild)): 1) spirits of madness; 2) herbalist's name for *Euphorbia ingens* (648); and
- **usukasihambe** (get out and let us go): medicinal plant sold by herbalists (768).

8   Names referring to the use of plants as charms

The blinkblaar [*Ziziphus mucronata*] is considered a magical tree by African tribes and believed to be an important tree to grow near the house as it will ward off evil spirits and lightning. It is also considered to be immune to lightning, so those who shelter under it in a storm will be safe. If the branches are cut after the first summer rains it is believed that a drought will surely ensue, so the wood may only be cut at certain times of the year (Roberts 1990: 32).

Introduction

In the quotation above, Margaret Roberts refers to a number of concepts relevant to the plants and plant names discussed in this chapter. For example, she refers to *Ziziphus mucronata* as a 'magical tree' and to the beliefs held by 'African tribes' that (1) the tree is 'magical'; (2) it will ward off evil spirits and lightning; (3) the tree itself is immune to lightning strikes; and (4) cutting branches of the tree after rain will result in drought.

This chapter is all about such beliefs and in the pages that follow we shall consider Zulu plant names that refer to

- charms used for evil purposes;
- protective charms of all sorts: protection against the malice of witches, extreme weather conditions such as lightning and hail, attacks by snakes, dogs, crocodiles and other wildlife, and bad luck;
- love charms and good luck charms;

- plants used as charms for positive and practical purposes; and
- the use of plants in various rites and rituals.

First, though, we need to consider the use of the word 'magic' and also discuss belief systems that use plants as protection against such a variety of threats.

'Magic'

In the opening quotation to this chapter Roberts is recorded as saying "The blinkblaar . . . is considered a magical tree by African tribes". Many writers about South African plants and their use by local communities use the word 'magic' regularly and without any comment. For example, let us look briefly at the comments made by Pooley and Boon about three species of tree. We start with *Ziziphus mucronata*, as the opening quotation in this chapter was about this tree. Pooley says "Leaves, roots and bark used for a wide range of medicinal and magical use" (1993: 298) and Boon echoes this with "Leaves, roots & bark used magically & medicinally" (2010: 336). As regards *Rhamnus prinoides*, Pooley has "Widely used for medicinal and magical purposes . . . as protection against lightning and evil spirits" (1993: 300), and again this is repeated in Boon (2000: 336) with "Used medicinally and magically . . . & as protection against lightning & evil spirits". Pooley is more brief when she says of *Kiggeleria africana* that it is "used for magical purposes" (1993: 328) and Boon far more brief with "used magically" (2000: 374).

Coates Palgrave tends to avoid the word 'magical'. He does say of *Ziziphus mucronata* that "The buffalo thorn has numerous magico-medicinal uses" (1977: 551), but for *Rhamnus prinoides* he merely says "Parts of the plant are widely favoured as a protective charm which is used against lightning . . . green twigs are burned to smoke away evil spirits from the fields" (1977: 555). The same 'narrative-descriptive' approach is followed in his comment about *Kiggelaria africana*: "The Tembu people believe that the tree should not be touched as it attracts lightning, but the South Sotho use parts of it for a potion to protect their kraals" (1977: 626).

As far as I can see, Hutchings never uses words like 'magic' or 'magico-medicinal'. Of *Ziziphus mucronata* she says, "Branches are used to attract ancestral spirits from an old dwelling site to a new one. They are also placed on the graves of chiefs and heads of kraals after burial" (1996: 193) (see Plate 22) and her comment on the use of *Rhamnus prinoides* is "Plants are used as protective charms against lightning and medicated pegs are placed in homesteads while green twigs are burned to protect lands from evil influences" (1996: 194). She does not mention *Kiggelaria africana*.

I have certain problems with the use of the word 'magic' in such descriptions of plant usage. The connotative meanings of this word in English are disturbing. For example, *Collins Dictionary of the English Language* gives the following definition:

> 1. The art that, by use of spells, supposedly invokes supernatural powers to influence events, sorcery. 2. The practice of this art. 3. The practice of illusory tricks to entertain other people, conjuring. 4. Any mysterious or extraordinary quality or power (1986: 925).

The word 'sorcery' in this definition leads us to the following:

> the art, practices, or spells of magic, esp. black magic, by which it is sought to harness occult forces or evil spirits in order to produce preternatural effects in the world (1986: 1456);

and 'occult' leads us to

> 1. of or characteristic of, magical, mystical or supernatural arts, phenomena, or influences; 2. beyond ordinary human understanding; 3. secret or esoteric (1986: 1064);

while 'preternatural' produces "beyond what is ordinarily found in nature; abnormal" (1986: 1213).

Again, the connotative meanings of many of the words used in these definitions are disturbing, such as 'supposedly', 'sorcery', 'illusory tricks', 'conjuring', 'mysterious', 'occult', 'secret' and 'abnormal'. The emphasis, especially in relationship to the word 'sorcery', is on 'black magic', a phrase heavy with negative connotations and associations. Let us look at what two very well-known South African writers on Nguni traditional practices have to say about 'magic'.

The renowned South African anthropologist W.D. Hammond-Tooke has some interesting and useful observations to make. Although writing about the Bhaca people of the Eastern Cape, his observations are pertinent to Zulu society. After pointing out that the "integration and smooth functioning . . . [of society] . . . are ensured by the mechanisms of political institutions and the supernatural sanctions of religion", he goes on to say:

Man has developed, however, a further powerful tool, based on culture and language, to bolster the foundations of his society and ensure its persistence and well-being. This is the system of magical beliefs which proclaims that man is able to control the powerful forces of nature and the ravages of sickness and death. The belief in magic plays an important integrating role by inculcating confidence in the society's ability to overcome destructive agencies (1962: 263).

We see here that rather than focus on notions of 'occultness', 'sorcery', 'illusory tricks' and abnormality, Hammond-Tooke sees a belief in magic as culturally and socially positive, bolstering the "foundations of . . . society" and ensuring its "persistence and well-being". He does however, acknowledge that there are those use who use 'magic' for negative purposes; to do the opposite of what is said above. He puts it like this:

But magic, the product of man's mind, has developed a pathology, a perversion of its techniques and aims into antisocial channels directed against society, and this, in its turn, has resulted in the development of magic to protect society against these misuses. In this chapter we shall discuss the nature and function of magic to approve socially approved ends; in the next its aberration in the activities of the witch and the sorcerer (1962: 263–4).

We would appear here to have magic functioning in three ways in society: <u>first</u>, used in a positive way to bolster the "foundations of . . . society"; <u>second</u>, in a negative way as "antisocial channels directed against society"; and <u>third</u>, in a positive way to counteract the negative intentions of those who use magic against society.

Let us now look at what Berglund (1976) has to say about both the use of magic and use of the word 'magic' in ethnographic description. Berglund notes that "in anthropological literature the word magic has been used extensively – and unfortunately also misused" (1976: 27). He draws our attention to a definition of magic given by Charles Winick in his *Dictionary of Anthropology*, namely that it is "the techniques of coercion, based on what we would call false premises, by which persons, usually non-literates, try to obtain practical ends" (1964: 354). It is not quite clear whether Berglund is objecting to "techniques of coercion", "false premises" or "usually non-literates" or perhaps to the whole of Winick's statement, but he is quite clear that if that is how Winick perceives magic, then that is "exactly what this study [i.e., Berglund's] tries not to do" (Berglund 1976: 27). Rather, Berglund favours the distinction put forward by Mary Douglas in her 1966 *Purity and Danger* between "ritual acts which contain a meaningful and appreciable symbolic interpretation" on the one hand and "acts which are magic because they lack interpretation" on the other (Berglund 1976: 27). "Adopting this view and relating it to this study, i.e., to describe intelligibly Zulu thought-patterns and symbolism", says Berglund, "the use of the word magic automatically falls away" (1976: 27).

For Berglund the manipulation of materia for the purposes of love charms, protection against evil, warding off lightning and so on are all acts falling into Zulu thought patterns that are both systematic and intelligible. Such acts cannot, therefore, be part of a definition that includes such concepts as 'illusory tricks', manipulation of 'occult powers', 'beyond human understanding' and 'abnormal'. For the purposes of his book, then, "the word magic automatically falls away".

Although Hammond-Tooke <u>does</u> use the word 'magic', it is clear that he and Berglund are on the same wavelength when it comes to describing the use of plants and other materia for love and protective charms as part of a logical and systematic system of beliefs:

> Much has been written on the psychological basis of the belief in magic; it is not intended here to recapitulate the theories which are accessible in the literature on the subject. It is sufficient to say man has created a logically self-contained system of beliefs and practices which supply him with techniques for mastering the forces of nature and controlling human destinies (1962: 264).

We quoted Hammond-Tooke earlier as saying that manipulation of materia for positive purposes has the function of "bolster[ing] the foundations of . . . society and ensur[ing] its persistence and well-being". He goes further in detailing these "social uses of magic" when he says:

> We are interested, for the purposes of this study, primarily in the sociological role of magic: in the part it plays in conveying authority and supporting the political structure of the tribe and the mystical attributes of the chieftainship, its importance in the economic sphere, in the individual life cycle (birth, initiation, marriage and death), and in intimate human relations (e.g. love magic). Then, too, the forces of the elements are often dangerous to man and magical techniques

are developed to protect man and beast against the devastation of lightning and hail. All these may be listed as the social uses of magic (1962: 263–4).

All of these will be illustrated in our discussion that follows of the underlying meanings of the Zulu names of plants used for these various purposes.

I would like to take the notion of 'intelligible, logical systems of belief' referred to by both Hammond-Tooke and Berglund and say that in the systems they talk about there is a 'three-way dynamic' involved. The three dynamic factors are:

- Human beings with their desire for a stable, ordered life (Hammond-Tooke's 'persistence and well-being of society');
- Threats to the stability and the order desired by humans; and
- Plants perceived to have powers that can be manipulated to protect against these threats.[1]

Included in the threats are the malicious intentions of *abathakathi* to disrupt stability and order; i.e., Hammond-Tooke's "antisocial channels directed against society".

Protective charms

Protection against evil

Earlier in this chapter, we noted, following the thinking of Hammond-Tooke, that 'magic' can function in three ways; namely positive, negative and to counteract the negative intentions of those who use magic against society. We shall start with a brief look at names that refer to the antisocial use of plants and then spend more time looking at names referring to use of plants to combat such antisocial usage. As Hammond-Tooke points out, such antisocial aims include sending sickness and death, harming fields and crops and causing stock to sicken and die. He further points out that not all such evil and malicious intentions come from *abathakathi*. He notes that a "complex of jealousies and emotional frictions, often unconscious" frequently develops among people living in close proximity to each other and says it is

---

1. I am quite aware that materia manipulated for various causes is also obtained from the animal and mineral kingdoms. However, as this book is about Zulu plant names, we shall only consider plant material.

significant that most allegations of witchcraft and sorcery are against those living in the same locality, even in the same homestead, and a high proportion of accusations are against a co-wife, neighbour or sexual rival. The polygynous marriage situation is a fertile field for accusation and counter-accusation, the psychological bases of which are readily understandable (1962: 278).

Given the situation outlined by Hammond-Tooke, it would seem that the average law-abiding and peace-loving citizen has to face evil and malicious intent from two directions: not only from the feared *abathakathi*, for whom negative actions against society are simply *raison d'être*, but also from apparently friendly and trustworthy neighbours and family members. Small wonder, then, that so many plant species are enlisted in protection against so many potential threats; and small wonder, also, that the lexicon of plant names makes such extensive reference to the need for such protection.

Many of the plants used as protection against evil have names based on a verb or ideophone with meanings such as 'block', 'render harmless', 'turn back' or 'dissipate', referring to their ability, if used properly, to neutralise evil intentions. The following are examples.

Doke and Vilakazi give the name **ingqunda** (< *qunda* (to blunt)) for "1) species of iris, *Moraea spathacea and *Homeria pallida, used as a palliative against witchcraft; and 2) any medicine used to blunt hostile action" (1958: 716). 'Moraea spathacea' is *Moraea spathulata* in Pooley, who gives the same Zulu name **ingqunda** and says "used in traditional medicine to treat infertility and as a protective charm" (1998: 236). Hutchings quotes Gerstner (1939) as saying that "preventative medicine made from the plant is used to 'blunt witches' deeds'" (1996: 59), showing clearly that the Reverend Gerstner was well aware of the relationship between the function of a plant and the underlying meaning of its name.

The verb *sula* (wipe off, out or erase) is the base of the name **insulansula**, which according to Doke and Vilakazi refers to a "protective charm against witchcraft placed in the kraal; charm for blinding,[2] made from the herb *Spermacoce natalensis*" (1958: 768). Hutchings gives **insulansula** as one of four Zulu names for *Spermacoce natalensis* and says, interestingly, that "leaves

---

2. Doke and Vilakazi actually say 'blinding', but in the witchcraft literature this kind of charm is usually referred to as a 'binding' spell'.

are chewed or put under the tongue after committing a fault as protective charms against the anger of teachers or judges" (1996: 301).

The name **idlula**, given by Pooley (1998: 408) for *Begonia homonyma* (used "for protective charms") is derived from the verb *dlula* (pass by, block something unwanted). Doke and Vilakazi (1958: 162) say that **idlula** ("a species of pink, flowering Begonia") is used as a preventive medicine. For Hutchings (1996: 209) root decoctions of **idlula** are taken to counteract *idliso* (poison believed to be administered in food). Similar sounding, but in fact derived from a different verb, is the plant name **ihlula**, from the verb *hlula* (overcome, beat, conquer). Pooley (1998: 328) gives this as a name for *Senecio oxyriifolius* and says it is used to prevent sorcery. Doke and Vilakazi do not recognise **ihlula** but Hutchings (1996: 329) gives it as a name for the same plant *Senecio oxyriifolius*, although she says nothing about its use as a protective charm.

Similar in underlying meaning is **isithibane** (< *thiba* (restrain, check, ward off)), a Zulu name for the Hairy Caterpillar-Pod (*Ormocarpum trichocarpum*) (Boon 2010: 162). Hutchings (1996: 140) says that emetics made from the bark are used for suspected poisoning.

Pooley has **iphinganhloya** (what suppresses the ill-omen) as a Zulu name for the Broom Asparagus (*Asparagus virgatus*), which is used as a protective charm to ward off the effects of evil (1998: 104). Doke and Vilakazi say of the name **iphinganhloya**: "wild asparagus, **A. plumosus*, used magically in warding off the effects of evil" [< *phinga* (cool down, quieten, suppress) + *inhloya* (obsolete form of *umhlola*, omen); i.e., what suppresses the ill-omen] (1958: 663). Doke and Vilakazi also give this as: **iphunganhlola** (what drives off bad luck): "species of asparagus" [< *phunga* (scare off, drive away) + *inhlola*³ (ill-omen)] (1958: 678). They also give **iphunga**, a name used for two species of shrub: *Senecio pleistocephalus* and **Vernonia podocoma* as well as for "the creeper, *Tephrosia kraussiana*, used as a medicinal charm for doctoring cattle against quarter-evil" (1958: 678). In this last-named function, the meaning 'whisk away flies' in the verb *phunga* may be relevant; that is, if the condition 'quarter-evil', whatever that might be, is caused by flies.

---

3. Note that Doke and Vilakazi give three forms of the word for 'ill-omen': *umhlola*, the form most commonly encountered today; *inhlola*, which is simply the same noun stem in a different noun-class; and *inhloya*, the 'thefuya' form of *inhlola*. 'Thefuya-speech' is a characteristic of the now-defunct Lala dialect of Zulu, where a final 'l' was pronounced as 'y'.

Bryant gives us some very interesting information that links the plant **iphinganhloya** and threats to well-being. Under a discussion of the word *izembe* (axe, hatchet) with secondary meanings that relate to diseases suffered only by men, and the medicines to cure them, he says:

> a man who, in battle or otherwise, has killed another contracts (potentially, though not actually) a similar disease; and in order to clear himself of it, he must, before re-entering his kraal, go through an elaborate course of purification . . . He must carry on his head a spray of the *i(li)-Pingantloya* plant, must treat himself with all manner of herbs, must abstain from several kinds of foods, and above all must remain and sleep out on the veldt, until he has had sexual intercourse with some strange female (or in case of necessity, any boy) he may chance to come across (*uku-sula izembe*) (1905: 725).[4]

Let us look at more plant names containing bad luck and evil omens. Hutchings gives **ugibisisila** as a name for *Bowiea volubilis*, which is "used as protective washes when travelling" (1996: 30). Doke and Vilakazi give this in the form **ungibisisila** (what drives off bad luck)[5] for the same species and say "used as protective and love charm" (1958: 555). They also give **ungibamhlola** (what wards off ill-omen) for a "species of medicinal plant" (1958: 555) and **isigibamhlolo** (< *giba* (take out) + *umhlolo* (what wards off an evil portent)) for a "species of plant" (1958: 245). Note yet another form of the word for 'evil omen'.

Counteracting witchcraft and malice can be seen as a prospective victim hitting back and the verb *shaya* (to hit) is used in a couple of names of plants used for this purpose. The name **umphindamshaya** (hit him again) is given by Boon for *Adenia gummifera*, which he says is "used in counteracting witchcraft" (2010: 384). Doke and Vilakazi give this name in the forms **uphindamshaye** and **impindamshaye**, which they say refers to a "species of wild grenadilla, *Adenia gummifera*, used as a medicinal charm to react on a witch" (1958: 662). They explain the derivation as from the verb *phinda* + the subjunctive form of the verb *shaya*, with the underlying meaning 'what turns back that he may strike him'. Hutchings also gives the simpler form **impinda** (what turns

---

4. *ukusula izembe* (to wipe out the axe): in other words by wearing the *iphinganhloya* sprig on the head, abstaining from foods and sleeping with stray women or boys a man removes the potential of the various 'izembe' diseases.
5. The word *isisila* means 'bad fortune, ill luck, disfavour, stigma'.

back) for this plant, and says "sprinkling protective charms are made from the plant" (1996: 208).

Also using the verb *shaya* is **ushayabhici** (what strikes and destroys), a name Pooley gives for the plant Christmas Bells (*Sandersonia aurantiaca*), used as a "charm for protection against evil" (1998: 26). Doke and Vilakazi (1958: 734) say that **ushayabhici** is derived from the verb *shaya* and the ideophone *bhíci* (of destroying). Their primary meaning for this word is "wet leprosy [?] or spreading epidemic", while the second meaning refers to a plant for which they give the botanical name '*Saundersonia* [*sic*] *aurantiaca*' and the English vernacular name 'Christmas Bell Lily'. Surprisingly, although Hutchings (1996: 27) gives three Zulu names for this plant, she does not include **ushayabhici**.[6]

When not actually deflected or turned back, evil intentions can be 'tricked' or 'puzzled' as in the name **iphamba** that Pooley (1998: 128, 130) gives for *Diaphananthe millarii* ("used as a protective charm") and *Cyrtorchis arcuata* ("used in traditional medicine as a love potion"). Doke and Vilakazi say that **iphamba** refers to both "mistletoe; a parasitic plant of the orchid family" and a "species of climber of the periwinkle family, **Oncinotis inandensis*, used to counteract witchcraft" (1958: 644). Hutchings (1996: 247), who also gives **iphamba** as a Zulu name for *Oncinotis inandensis*, quotes Palmer and Pitman (1972) as saying this plant is used to counteract witchcraft, as medicines made from it are believed to be powerful enough to confuse sorcerers. It is perhaps interesting to note that the verb *phamba* is also used in the name of the KwaZulu-Natal South Coast river the uMphambanyoni (what puzzles the birds), a river that twists and meanders so much that birds flying over it are continually puzzled by its apparent changes in direction of flow.

The idea of puzzling, misleading or confusing those with evil intentions can also be seen in **impundu** (gatepost) a name Pooley (1998: 342) gives for the Variegated Aloe (*Gasteria croucheri*). Doke and Vilakazi give the meaning of this word as both 'gatepost' and "Species of plant, **Gasteria glabra*, whose bulbous roots are placed between the gateposts at the kraal entrance to cause forgetfulness in evil-doers" (1958: 677). They do not say so, but this word for 'gatepost' is surely related also to the word *impundulu*, according to Doke and Vilakazi a "bird supposed to be used by women in witchcraft" (< *phundula*

---

6. But she does give Chinese Lantern Lily as an English vernacular alternative for Christmas Bells.

(lead astray, mislead, puzzle, confuse)) (1958: 513). The word *impundulu* is also used for the fabulous 'lightning bird' and we shall return to this bird, and this word, when we discuss protective charms against lightning.

The Zulu name **umayime** is used for a number of plants (see Plate 15). Pooley (1998) lists *Clivia gardenii*, *Clivia miniata*, *Osyridicarpos schimperianus*[7] and *Mentha aquatica* with this name, while Hutchings (1996) assigns the name to an unidentified *Brunsvigia* sp., *Clivia miniata*, *Clivia nobilis* (quoting Gerstner (1939) as saying "Used in protective sprinkling charms known as *intelezi yezulu*") and *Mentha aquatica* ("infusions are used as sprinkling charms against evil spirits"). The name **umayime** means 'let it stop' (< *ma* (let) + *yi* (it, which could refer to any noun represented by the concord 'yi'; perhaps *inhlola* (evil omen)) + the subjunctive form of *ma* (stand, stop, halt)). For Doke and Vilakazi the name refers to the "Bush lily, St. John's lily, *Clivia* species" (1958: 489) whose roots are used for snakebite.

Bringing to a stop or halt is also expressed in the Zulu verb *bopha*, the base of the name **umabophe**, that Boon gives as a name for the Moth-Fruit (*Acridocarpus natalitius*), saying that its "roots [are] used . . . magically . . . as *intelezi*, a name given to all charms and plants used against evil" (2010: 224). Hutchings says of the same plant that "roots may be chewed or put under the tongue as a charm to avert anger when a fault has been committed" (1996: 159). She also notes its use as sprinkling charms to avert danger and as protective war charms. Doke and Vilakazi (1958: 474) give this name for both *Acridocarpus natalitius* and *Spermacoce natalensis* and indicate that it is the latter plant whose leaves are put under the tongue when a fault has been committed. As we saw earlier when discussing the name **insulansula**, Hutchings records this use of the leaves of *Spermacoce natalensis* as well, so both these plants are used to cover up verbal indiscretions.

The Zulu plant names **umlalume** and **inyazangoma** give rise to a number of interesting questions. The first is one of three Zulu names Boon (2010: 130) gives for *Prunus africana* (African Almond or Red-Stinkwood). The name means literally 'what lies down and gets up again'. Doke and Vilakazi say of the name **umlalume** that it is a "tree of the Pygeum[8] species, used in treating witchcraft" (1958: 447). Clearly this plant is not so much used to <u>prevent</u> witchcraft, but to help one <u>recover</u> after being bewitched.

---

7. *Osyridicarpos schimperianus* also has the name **umalala** (let it lie down).

8. *Pygeum africanum* is an earlier synonym of *Prunus africana*.

Although Hutchings (1996: 118) does not give **umlalume** as a Zulu name for *Prunus africana*, she does give **inyazangoma** (actually **inyazangoma-elimnyama** (black inyazangoma)) and this is a Zulu name with a number of potential interpretations, all possibly, but not convincingly, related to Doke and Vilakazi's "used for treating witchcraft" (1958: 447). Doke and Vilakazi (1958: 623) do give the name **unyazangoma**, but offer no suggestion as to derivation and say rather vaguely "Two species of trees; of one the roots are used medicinally against dreams; the other is an *Albizia*, a leguminose of the Mimosa type" (1958: 623). I am reluctant to see this name as a compound of *unyazi* (lightning) and *ingoma* (drum, ritual song) and think it more likely to be a compound of *-nya* and *izangoma* (diviners). The problem is that *-nya* could be any of the following (all from Doke and Vilakazi (1958: 616): (a) the ideophone *nyá* (of nothingness, silence, ending); (b) the verb *nya* (to excrete); or (c) the noun stem *-nya*, which occurs in the following three noun classes:

- *i(li)-nya*: a class 5 noun with a number of apparently unrelated meanings: 1) desire to exert oneself; 2) feeling of vengeance; 3) cast-off smeltings; 4) belly or underparts of snake; and 5) private or unpleasant things;
- *u(lu)nya*: a class 11 noun meaning 'cruelty, callousness, hardness of heart'; and
- **umunya**: a class 3 noun meaning "species of evil-smelling plants, *Papaver somniferum* and *Cannabis sativa*".

Seeing that Hutchings says of *Prunus africana* that it "smell[s] strongly of bitter almonds" <u>and</u> "fruit is reputed to be used in witchcraft" (1996: 118), it is very likely that in the name **inyazangoma**, the element *umunya* (evil-smelling plant) is combined with an element meaning 'diviners' for it is, after all, diviners who are largely responsible for treatment after witchcraft.

Haworthia limifolia (Pooley 1998: 342: no common name) has two names relating to different aspects of protection: **isihlalakahle** (living well) refers to the peace that comes from protection; while **umathithibala** (being helpless) is what happens to intruders when they run into this plant. Hutchings (1996: 36) says whole plants are used as sprinkling protective plants (*intelezi*) and quotes Cunningham as saying that plants are cultivated in containers on hut roofs as a protection against lightning. The closely related *Haworthia fasciata* (also with the name **umathithibala**) appears to be focused specifically against the mischief of *utokoloshe* (see Plate 18).[9]

The Swazi Acacia (*Acacia swazica*, Boon 2010: 186) has the Zulu name **ukhalimele** derived from the verb *khalima* (turn back someone, warn off, prevent dangerous movement). Hutchings assigns this name to "unidentified *Rhynchosia* species" (1996: 145) and says they are used as love charms, which does seem rather at odds with the underlying meaning of the name. Clearly she gets this from Doke and Vilakazi who give the first meaning of the name as "forest creeper of the genus *Rhynchosia*, used for headache and as love charm". Their second meaning is "*Acacia swazica*, a shrubby tree of the northern bushveld" (1958: 376), showing that while Boon (or, rather, Pooley (1993) before him) has gone for Doke and Vilakazi's second meaning of **ukhalimele**, Hutchings preferred the first.

Amangwe ezidlayo, a name of the Greenleaf Wormbush (*Cadaba natalensis*, Boon 2010: 114) needs explanation. The noun *amangwe*, always used in the plural, refers to any kind of medicine used for evil and destruction, while *ezidlayo* literally means 'which eats itself'. Taking protective medicines from this tree, then, has the effect of turning back any dangerous medicines and redirecting them to the person who sent them in the first place. You could say that this name means 'destructive medicine that self-destructs'. Of interest here is Hutchings' statement of *Cadaba natalensis* that "Roots are used . . . to <u>cure</u> pleurisy believed to be caused by witchcraft . . . They are also believed to be used in witchcraft to <u>cause</u> pleurisy [my emphases]' (1996: 111). Both Boon and Hutchings give the alternative name **amangwe amnyama** (black amangwe) for this plant. Bryant gives interesting details about how any *amangwe* medicine should be administered:

Umtakati awupeke umuti odengezini, awuncinde, agcobe ngawo imikhonto emibili; ab'es'eyiposa leyo'mikonto ngakuye lowo'muntu atand'ukumbulala, angabe esabeka ngakona, lowo'muntu ab'es'efa njalo amanxeba nokukwehlela.

(Let an *umthakathi* cook the medicine in a potsherd, scoop it up with (lit. 'suck it off from') the fingers, and then smear it onto two spears;

---

9. *Utokoloshe* (or *utikoloshe*) is a "wicked little dwarf who lives in deep pools or in the reeds. He is short and hairy, and very fond of women" (Krige 1950: 354).

and then fling the spears at that person whom he wishes to kill, and while he simply looks on, that person will die there and then from wounds and coughing) (1905: 430).

The Lavender Fever-Berry (*Croton gratissimus*, Boon 2010: 234) is a tree that protects against mortal threats, as evidenced by the name **um(a)hlabekufeni** (< *ma* (characteristic) + *hlaba* (stab) + locative form of *ukufa* (death; i.e., what stabs at death)). Although *ekufeni* is derived from *ukufa* (death), Doke and Vilakazi translate this name as "what strikes at sickness" (1958: 477). Hutchings (1996: 166) gives an extensive list of the medicinal uses of this tree. One is that ground bark is rubbed into incisions on dropsical swellings and this may be related to another name Hutchings gives for this species: **intumbanhlosi**. The Zulu word *ithumba* (tumour, abscess; cf. *intumbane* (boil)) is easily recognisable in the name **intumbanhlosi**, but what is the second element: could it be *inhlosi* (sea barbel); or perhaps *ihlosi* (type of leopard larger than normal)? Both are distinctly unlikely. It is possible that -*nhlosi* is derived from the verb *hlosa* (aim at, target), making **intumbanhlosi** a 'medicine targeted against boils and tumours'.

Similar in meaning and intention to **umahlabekufeni** is **usahlulamanye** (what overcomes the others), a name for the White Candlewood (*Pterocelastrus echinatus*, Boon 2010: 312) although Hutchings (1996: 185) mentions its use only for respiratory ailments and digestive tract and blood-cleaning purgatives. She also gives the name **usahlulamanye** to *Pterocelastrus rostratus* (the Red Candlewood), with much the same medicinal uses.

Another plant name suggesting general protection from unspecified threats is **umhlangula** (< *hlangula* (remove a danger from)), which refers to the Magic Guarri (*Euclea divinorum*, Boon 2010: 466). Hutchings (1996: 232) gives a number of medicinal uses and states that Venda men use this plant if they have slept with a woman made pregnant through adultery. There is no indication in the literature that Zulu men use the plant in the same way.

The word *inkungu* (mist) appears to be used in a metaphorical manner in the neutralising of evil. Pooley gives **inkunkwini** as a Zulu name for *Plectranthus hadiensis*, which she says is used "as a sprinkling charm against evil spirits" (1998: 186). Doke and Vilakazi give the more reliable spelling **unkungwini** (< *enkungwini* (in the mist) < *inkungu* (mist)) and say this is a "species of *Plectranthus* used as a love charm" (1958: 580). Whether the reference is to malice and evil, or to reluctance towards a suitor, clearly the medicines made from the plant make these less focused, and misty. Hutchings (1996: 289) gives the name **unkungwini omncane** (small unkungwini) for *Crabbea nana* and says it is used as a protective sprinkling charm against storms, so we shall discuss it in more detail under the heading dealing with protective charms against adverse weather.

More references to the warding off of evil can be seen in the Zulu names for *Rapanea melanophloeos*, which Boon says is used as an "*intelezi*, a charm to ward off evil" (2010: 446). He gives the two names **isiqalaba** and **umaphipha**. Of the first Doke and Vilakazi say "**isiqalaba**: species of forest tree, *Myrsine melanophleus* [= *Myrsine melanophloeos*] (< *qalaba* 'be self-confident')" (1958: 686); and of the second "**umaphipha**: herbalist's name for the tree *Rapanea melanophloeos* (< *phipha* 'help through difficulty, assist one in trouble')" (1958: 485).

Pooley gives the name **udambisa omkhulu** for *Xysmalobium involucratum* and says "used as a sprinkling charm against evil" (1998: 166). Doke and Vilakazi say **idambisa** is derived from the verb *dambisa* (cause to subside) and refers to a "small veldt shrub, the leaves of which are used for rendering boiling water ineffective" (1958: 138). They also give **udambiso** as a "species of wild vine used for poulticing, **Cissus connivens, *C. lanigera*"). Clearly, whether this shrub and/or vine is used for sprinkling against evil, rendering boiling water ineffective or as a poultice, the underlying meaning of 'cause to cool down' is relevant.

Pooley has the name **ubuhlungubendlovu** (elephant's pain) for the Sand Forest Poison Rope (*Strophanthus petersianus*) and says it is used "as a charm against evil" (1998: 166). Hutchings confirms this with "unspecified parts are used by herbalists as a charm against evil" (1996: 248), but neither author offers any clue as to the relationship between the use of the plant and the underlying meaning 'elephant's pain'. Doke and Vilakazi (1958: 338) translate **ubuhlungubendlovu** as 'elephant's herb' and say it is used to treat hysteria, but that gets us no closer to understanding the name.

It would seem from the meaning of the verb *finga* (render harmful) that any plant names based on this verb would indicate a plant used by evil-minded persons to cause harm, rather than to protect against harm. And yet Doke and Vilakazi give us the noun **imfingo** with various meanings: "1) medicine used for counteracting a harmful charm; 2) **Stangeria paradoxa*, a small Cycadacea . . . used as a counteracting charm; and 3) a species of climbing fern" (1958: 207). From the noun *imfingo* is derived a further noun: **imfingwane** (a small species of *Stangeria paradoxa* fern).[10] This apparent paradox is explained by the fact that several species of plant may be used both to cause and to cure certain undesirable conditions. As we noted before in Chapter 7 on traditional medicines, such plants are unlikely to be the only ones used in a concoction and the role of the catalyst is all-important. If an *umthakathi* or other evil-minded person were to use the plant *imfingo*, the catalyst would be an <u>activator</u>; while if the plant were used in medicines designed to counteract witchcraft, the catalyst would be a <u>negator</u> or <u>neutraliser.</u>

Threats to well-being are not only the results of directed malice. Soldiers at war face distinct threats to their well-being and it has been well documented that soldiers in the Zulu armies were doctored by traditional healers before setting out to fight. Two plant or medicine names refer to this specifically: Doke and Vilakazi give the name **umpikayiboni** (the army does not see) and say this refers to a "medicine to make people invulnerable" (1958: 511). It is also Doke and Vilakazi (1958: 797) who give the plant name **umuthi-wamadoda** (medicine of men) to refer to the Leadwort (*Plumbago capensis*), used as a warrior medicine.

Another name refers to a plant used for doctoring soldiers, but with many other uses as well: Doke and Vilakazi state that the plant name **impengu** is derived from the ideophone *phéngu* (of reversing, changing) and refers to "Species of shrubs of the *Capparis* family, *Maerua woodii*, *M. triphylla*, *M. angolensis*, *M. rosmarinoides*, [and] *Cadaba natalensis*, used for smoking the fields and for purifying soldiers" (1958: 656). The name is also used for a small divining bone. For Pooley (1998: 256), however, the name **impengu** refers to *Senna italica*. She says that in addition to being a charm used against evil spirits, it can also be used to treat pain, roundworm and as a snakebite remedy.

Use in witchcraft

Berglund says:

> Although the great majority of *imithi* are believed to be neutral in character, there are a limited number of medicines, prepared of certain poisonous materia (particularly vegetation) which work only evil (1976: 346).

---

10. Beaumont (personal communication) says this is not a fern. Apparently fern leaves and *Stangeria* leaves look similar, but they belong to different divisions. Doke and Vilakazi presumably copied the fern reference from Gerstner.

In Chapter 5, in a discussion about *abathakathi*, we discussed the tree *Synadenium arborescens*[11] (Zulu **umdlebe**), which goes under the ominous English name Dead Man's Tree. I quoted Coates Palgrave as saying:

> Africans are sure that this is one of the most evil of trees; they firmly believe that it cries like a plumed snake (which is said to bleat like a goat!) and that it lures people and animals towards it in order to kill them. They are convinced that the ground round the trees is white with the bones of slain animals and birds flying above will fall dead from the sky (1977: 454).

Coates Palgrave may have taken these details from Palmer and Pitman, or they may themselves have got their detail from the same source as Coates Palgrave, for they say of the same tree:

> The most feared of all trees belongs to Zululand. It is umDlebe, or Dead Man's Tree, sometimes known as the Crying Tree, *Synadenium cupulare*, with a highly toxic latex around which lurid legends have arisen. Zulus believe that the tree strikes a man dead if he approaches too closely; that the sight of it, or a wind that blows across it, may kill; that birds drop dead as they fly past it. Jobe, son of a witchdoctor, told his employer, Ian Garland that the tree bleated like a goat or like inDlondlo, the legendary viper with plumes on its back. He would not mind, he said, searching for it in his company because Garland had binoculars and they could look at the tree from afar (1972: 195).

Over 100 years ago, Blessing was writing in his field notebook about this tree:

> *Meget giftig, Hvis en fugl sættir sig i træct, dør den. En vildbuk, som hviler i den skygge, dør.*
> (Very poisonous. If a bird sits in the tree, it will die. A wild buck resting in the shadow of the tree dies) (Paulsen et al. 2012: 8).

Berglund adds to this picture of dread:

---

11. Beaumont states (personal communication) that *Synadenium arborescens* is a synonym for *Euphorbia cupularis*, which is in turn a synonym of *Synadenium cupulare*.

The **umdlebe** tree (*Synadenium cupulare*) is also known as the Dead Man's Tree and the Crying Tree. It is said to cry like a 'plumed viper' and be so poisonous that, like a mamba, strikes a man dead as soon as he approaches it (see pages 107 and 196) (drawing by Angela Beaumont).

> *Umdlebe* . . . is wholly vile. And its branches are used by witches
> in treating a corpse to become *umkhovu*. Not only is the smell of its
> flowers said to cause death, but any association with it is proof enough
> that the person in question is an *umthakathi* (1976: 346).

Berglund goes on to mention the tree *Croton sylvaticus* (Zulu **umzilanyoni**
(avoided by birds)) said to be so poisonous that even birds die if they sit on
the branches.

There are really two separate issues here: some plants are 'wholly vile', that
is, the evil resides in the plant itself, rather than in the way it is manipulated;
while other plants may be used for 'vile' purposes by *abathakathi* and other
malicious-minded people, but 'legitimately' by other people. Plants in this
latter category usually have a 'two-way' function, as illustrated by Berglund
below on the medicinal use of crabs' eyes:

> Illegitimately, crabs are used medically in *ukuthakatha* to cause the
> eyes to fall out because the eyes of a crab stick out. Legitimately, crabs
> eyes are used to draw out the eyes of a person who has them deeply
> set in the head, a characteristic which is looked upon as ugly and is
> ridiculed (1976: 357).

Many plants are used for purposes of witchcraft and sorcery, but few plants
have this use reflected in the name, probably because as in *ubuthakathi*
(sorcery) itself, the use of such plants must be kept hidden. An example of a
plant name that illustrates the link between plant and evil is **ihlaba**, for which
Doke and Vilakazi (1958: 308) give three meanings: 1) sharp pain; 2) Prickly
Aloe (smaller than *umhlaba*), used in dressing hides; *Sonchus dregeanus* Milk-
Thistle; 3) soil from a grave, said to be used in witchcraft to cause lung disease
when mixed with food.

Before we go on to other plant names, let us look at a name for an
umthakathi. As we have seen above, the *umthakathi* is against the perpetuation
of the norms of society and procreation that ensure stability through generations.
This is seen clearly in one of several names for an *umthakathi* in the Zulu
language: *isalakwanda* (< *ala* (refuse, negate) + *ukwanda* (increase; i.e., one
who refuses increase in a family)) (Doke and Vilakazi 1958: 5).

We saw above that Berglund says of **Synadenium arborescens* (**umdlebe**)
that its branches are used by witches to treat a corpse to become *umkhovu*. The
word *umkhovu* is used in two names for various species of cycad: **isidwaba-**

somkhovu (the *umkhovu*'s leather skirt) and **isigqiki-somkhovu** (the *umkhovu*'s wooden headrest) (see Plate 23).

To understand the connotations of these two names one needs to delve a little into the realm of *imikhovu* (the plural of *umkhovu*). These fearsome beings are among the most dreaded of the familiars of the *umthakathi*. Berglund (1976: 279) gives us three 'grades' of this familiar:

- first, they can be created by an *umthakathi* by exhuming the buried corpse of an adult, usually a male. This corpse, still 'alive' in the sense of being an *idlozi*, is now put to death as a spirit as well by driving a stake of the feared *umdlebe* tree[12] through the fontanelle until it comes out at the anus;
- second, *abathakathi* can exhume the corpses of children, bring them to life and then let them grow <u>bigger</u> by feeding them on the flesh and sexual organs of their kin; and
- third, *imikhovu* can be created by raising the issue of copulation between an *umthakathi* and a captive baboon.

The phrasing we read earlier from the pages of *Collins Dictionary of the English Language* come to mind here: "the art, practices, or spells of magic, esp. black magic, by which it is sought to harness occult forces or evil spirits in order to produce preternatural effects in the world" (1986: 1456).

Bryant's *Dictionary* adds interesting suggestions to the Zulu name **isidwaba-somkhovu:**

> as [the *umkhovu*] moves about in a kraal at night, it produces a noise resembling the smacking or rustling of a woman's leathern kilt . . . whence it is frequently dubbed *isi-Dwabakazana* i.e., a little old bit of a kilt (1905: 322).

Bryant explains how an *umthakathi* uses the plant **uqhume** (**Hippobromus alata*):

> An *umtakati* makes an infusion of the root of this plant, mixing therein a little earth from the footprint of a person he may wish to kill. Taking the mixture as an emetic, he vomits the whole into the hole of a snake, calling out the name of the particular person after doing

---

12. Discussed above.

so. The desired effect will be the speedy demise of the individual so conjured! (1905: 548)

Doke and Vilakazi (1958: 704) say that the plant name **uqhume** is derived from the verb *qhuma* (burst, explode) adding an extra dimension to the "speedy demise of the individual so conjured" mentioned by Bryant!

The Glossy Sourberry (*Dovyalis lucida*) has the name **idungamuzi** (what disturbs the homestead) and Ndela Ntshangase says it is used by witches to sow strife in a family.[13] This name is shared by the Natal Guarri (*Euclea natalensis*, Pooley 1993: 400), which also has the semantically similar name **ichithamuzi** (what disperses a homestead). Materia from these trees is mixed with other ingredients to form a concoction that when given to the inmates of a homestead under one guise or another causes them to argue and fight. The Thorn Pear (*Scolopia zeyheri*, Pooley 1993: 330) is known as **idungamuzi lehlathi** (forest dungamuzi).

I suspect that the adjective 'black' (Z. -*mnyama*) in **amangwe amnyama**, a name Boon (2010: 114) gives for *Cadaba natalensis*, does not refer to dark leaves but to the purpose of the user of materia from this tree, for the noun *amangwe* on its own means 'medicine used for evil purposes'. Another name for this tree is **amangwe ezidlayo**, already discussed.

The adjective -*mnyama* is also used in the name **umanzamnyama** (black/dark waters), which Pooley (1998: 138) assigns to the Giant Wild Anemone (*Anemone fanninii*). Doke and Vilakazi say that **umanzamnyama** is a "herbalist's name for the *Anemone caffra* whose roots are used to produce hatred" (1958: 485), while Hutchings says of *Anemone fanninii* that "the [burnt] black roots are used 'to make other people black' [indicating sorcery]" (1996: 98).

Doke and Vilakazi give the name **iphamaphuce** for a "species of weed, **Eclipta erecta,* used in witchcraft preparations" (1958: 644) and say the derivation is [*pha* + *mu* + subj. *aphuca* (what gives and takes away)]. Pooley assigns the name to *Eclipta prostrata*, and says "used . . . for sorcery" (1998: 216).

Both **ililo elikhulu** (great cry) and **uzililo** (wailings, mourning) are names Pooley gives for *Stapelia gigantea*, which she says is "used in sorcery, reportedly capable of causing death" (1998: 302). The English vernacular name – Giant Carrion Flower – underlines the sense of death.

---

13. Personal communication.

Pooley says of the Assegai Tree (*Curtisia dentata*) that the "bark [is] used in witchcraft preparations" (1993: 384) and this is reflected in the name **umlahleni**, which Doke and Vilakazi (1958: 445) translate as 'throw ye him away' (see Plate 21).

Evil purposes are also suggested in the name **inhlungunyembezi** (poison tears), a name Pooley (1993: 428) gives to the deadly poisonous Round-Leaved Poison-Bush (*Acokanthera rotundata*).

Boon (2010: 360) gives **umthelelo** as one of the Zulu names of *Ochna arborea*. Doke and Vilakazi (1958: 790) say this word is also used to refer to a poisonous concoction placed on the ground to kill by witchcraft.

Doke and Vilakazi give the name **ushayakhuhle** (what strikes blind) for a "species of tree with red sap, used as a harmful charm" (1958: 734). They provide a translation of this name as 'what strikes blind', but in fact the second half of the noun is the ideophone *khúhle* with the meanings 'of thick, heavy darkness' and 'of destroying, fighting'. Both emphasise the evil nature of witchcraft.

To end this section on plants used for evil purposes, let us consider the name **umaphekemoyeni-omnyama**, which Boon (2010: 554) gives for *Oxyanthus latifolius*. The name is derived from *ma* (characteristically) + *pheka* (cook) + *emoyeni* (in the wind, in the air) + *omnyama* (which is black; i.e., that which is cooked in the dark wind). Boon says nothing about antisocial uses and the name is not recognised by the other writers I have consulted regularly, so the origin of this name is somewhat of a mystery. There is no doubt, though, that there is something of the dark occult about it.

Protection against the weather

As we have noted earlier, the noun stem *-zulu* occurs in a number of noun classes, most referring to certain aspects of the Zulu people (*isiZulu*, Zulu language; *amaZulu*, Zulu people; *ubuZulu*, Zulu culture, etc.) The base word for all these, however, is the class 5 noun *izulu* with the two related meanings 'sky' and 'weather'. Indeed, most weather comes from the sky and very often dangerously so: destructive winds, hailstorms, snow, torrential flood-causing rains, and thunder and lightning.

We see the word *izulu* in the name of an unidentified plant found in Doke and Vilakazi: **inyenyezulu**, for which they give the underlying meaning 'what deals stealthily with lightning' and say it refers to a "species of plant, used in lightning protection" (1958: 625).

Pooley notes of the *Agapanthus* genus, indigenous perennial herbs in the KwaZulu-Natal region, that they are excellent garden plants first grown in Holland in the 1680s. Of *Agapanthus praecox* she notes that this is a "popular garden plant, in cultivation since 1680s" and that it is used "for protective charms" (1998: 450). There is no separate comment on the Zulu name given – **ubani** – a name *Agapanthus praecox* shares with *Agapanthus campanulatus* (see Plate 19). The Zulu name for this plant is also a word in the Zulu language for lightning and one of the major uses of this plant in Zulu society is its cultivation as a protective charm against lightning (Hutchings 1996: 36, 37; Doke and Vilakazi, 1958: 67).

Lightning is one of the most feared threats to the stability of Zulu society as it comes suddenly, with great noise, and often leaves multiple deaths in its wake. As I write this in March 2011, I have in front of me yesterday's newspaper[14] with a headline reading "Lightning kills five members of one family". The story notes that a 37-year-old mother died with four children in her arms in eMahlwazini village in the Bergville region. Over the previous four months (the summer of 2010 and 2011) similar reports were found in the same newspaper two or three times a month. A close reading of Pooley (1998) and Boon (2010) shows that dozens of plants are used as parts of medicines used to protect against lightning and we shall look at those plants with underlying names that reflect this usage. At the same time we shall look at similar references in plant names to their use against hail and other adverse forms of weather.

That lightning is both feared and greatly respected (if this is not a contradiction of terms!) can be seen in the name **umhlonishwa** (that which is respected, from the passive form of *hlonipha* (show respect to)). Pooley gives this as a name for *Phymaspermum acerosum*, of which she says "used as . . . charms to ward off lightning" (1998: 320). Doke and Vilakazi say this name refers to a "small shrub, *Psoralea pinnata*, with bright blue flowers, burned to ward off lightning" (1958: 335). For Hutchings (1996: 136, 325) this Zulu name refers to *Psoralea pinnata* <u>and</u> *Phymaspermum acerosum*. Hutchings makes no mention of use as protection against lightning for *Psoralea pinnata*, but for *Phymaspermum acerosum* she notes that "unspecified parts are used as charms to ward off lightning".

---

14. *The Witness* 11 March 2011: 3. And then on 14 March, *The Witness* had the headline 'Durban: Two killed by lightning'.

Another name with a fairly overt underlying meaning is the Swazi *licishamlilo* (quench the fire), which Pooley gives for the White Paint Brush (*Haemanthus albiflos*) she says is used "as a protective charm against lightning" (1998: 104).

A tree with a Zulu name telling of protection against lightning is the Cape Teak (or Cape-Teak Bitterberry, *Strychnos decussata*). Boon gives as one of several names for this tree **umkhombazulu** (what points to the sky) and says of it that it is "used with crocodile fat for protection against lightning" (2010: 482).

According to Doke and Vilakazi (1958: 666), the Zulu ideophone *phíshi* (of fizzling out, being ineffective; of darkness, blackness) is the base of the plant name **impishimpishi** that Pooley (1998: 260) gives for *Aspalathus chortophila* used traditionally, she says, for protection against lightning.

The name **umbophe**, which, as we saw earlier is derived from the verb *bopha* (halt, stop, arrest) and refers to plants used to halt witchcraft, may also refer to plants used against lightning. Pooley gives it as a Zulu name for *Conostomium natalense*, which is "used traditionally as love charm and to prevent lightning" (1998: 492).

Pooley gives **isihlalakahle** as a name for *Haworthia limifolia*, "used in traditional medicine as a protective charm, particularly against lightning" (1998: 342). Pooley explains that plants in the *Haworthia* genus are "small . . . like miniature aloes", which certainly contradicts Doke and Vilakazi's entry for **isihlalakahle** (< *hlala* (live) + *kahle* (well, safely)) as a "species of tree,[15] *Haworthia limifolia*" (1958: 314). Hutchings (1996: 35, 36), who confirms Pooley's description of the *Haworthia* genus by referring to them as "small succulent perennials", gives the name **isihlalakahle** for *Haworthia limifolia* and refers to a personal communication from Tony Cunningham[16] when she says that "whole plants are cultivated in containers on hut roofs as protective charms against lightning". She also gives **umathithibala** for this plant, a name used for a number of plant species involved in protection of various kinds (see Plate 18). For example, Pooley gives this name for *Aloe aristata*, used "as a protective charm against lightning" (1998: 32). This is probably one of the best-known Zulu medicinal plants, seen planted outside almost every Zulu

---

15. My emphasis.
16. An ethnobotanist at the University of Natal in the 1990s who did a considerable amount of fieldwork and research into plants used by Zulu people.

homestead. Doke and Vilakazi (1958: 796) say the name is derived from the verb *thithibala* (be helpless, low-spirited, depressed).

We saw the use of the verb *nqunda* (blunt) used in the names for **Moraea spathacea* and **Homeria pallida*, both "used as a palliative against witchcraft" according to Doke and Vilakazi (1958: 716). Pooley gives the name **umnqunda** for *Kohautia amatymbica* as well, saying that the plant is used as "protection against snakes, lightning" (1998: 202).

From the ideophone *bhóqo* (of subsiding, giving in) comes the Zulu plant name **ubhoqo** given by Doke and Vilakazi (1958: 45) for several species of *Ipomoea* used as love charm emetics. Pooley gives the name **ubhoqo** for *Ipomoea pellita* and says "used as love charms and protective charms against lightning" (1998: 422). It is remarkable, in fact, how many plants appear to be used both for love charms and as protection against lightning. The Wild Ginger (*Siphonochilus aethiopicus*), on the other hand, is used for protection against lightning and also against snakes, although presumably not at the same time. Pooley (1998: 360) gives the names **isiphephetho** and **indungulo** for this plant, while Doke and Vilakazi (1958: 175) use **umphephetho** (< *phephetha* (blow powdered medicine in air)) and **indungulu** (no derivation suggested). They call the plant Natal Ginger.

For the succulent-leaved herb *Talinum caffrum*, "used as . . . a charm against lightning" (1998: 250), Pooley gives the two Zulu names **inkucula** and **umpunyu**. Doke and Vilakazi (1958: 408) say the first name is derived from the verb *khucula* (wipe away, sweep away, remove, destroy, wipe out). Although they agree that **inkucula** is a name for *Talinum caffrum*, they give **uphunyu** as referring to "*Portulaca* herbs used as preventive medicine" (1958: 678) and the derivation of the name from the ideophone *phúnyu* (of slipping free).

Hutchings gives the Zulu names **isiphemba** and **isiqumaṉa** for *Xerophyta retinervis* and says "used as a protective medicine against lightning" (1996: 56). Doke and Vilakazi say **isiphemba** refers to a "species of shrub **Vellozia retinervis*, used as lightning medicine" (1958: 654) and that the name is derived from the verb *phemba* (light a fire, kindle). Clearly, as with a number of examples discussed above (and in the following entry as well), when this plant is used in protective medicines against lightning, a negator must be added as catalyst. Doke and Vilakazi give **isiqumaṉa** for the same plant and say it is "used as lightning medicine and for rope-making" (1958: 715). There would appear to be no underlying meaning in this name.

Pooley gives **isiwisa** (make fall) as a name for *Centella glabrata*, "used . . . as a charm against hail" (1998: 540). Doke and Vilakazi (1958: 853) give three meanings for the word, of which the second is 'hail' and the third 'medicinal charm to keep hail from the gardens'.[17] Here again, the underlying meaning of the name of the plant appears to be at odds with its stated function in the literature, and once again, as we have noted many times previously, the use of the correct catalyst is important. If you, jealous of your neighbour's healthy crops and livestock and holding a grudge against him for past injuries (whether real or imagined), hire me, an accomplished *inyanga yezulu* (sky-doctor), with few social morals and a great love of money, to cause hail to fall on the afore-mentioned neighbour's crops and livestock, I will be sure to include the plant **isiwisa** (cause to fall) in my concoction, and will add a catalyst that activates whatever it is in the **isiwisa** plant that causes hail to fall. On the other hand, if you hire me to <u>protect</u> you against hail, whether the result of the caprice of the heavens or directed by a jealous neighbour, I will use exactly the same ingredient, **isiwisa**, but include a catalyst that is a negator.

Hutchings' entry for *Crabbea nana* gives **ubuhlungu-besigcawu** (the pain of the open yard) and **unkungwini omncane** (little bit of mist): "used for snakebite and in protective sprinkling charms against storms" (1996: 289). One is tempted to see these two names each referring to one of the two specific uses mentioned: the pain in the yard comes from the bite of a snake and the 'little bit of mist' is what you hope a storm will turn into when this plant is applied to challenge it. Doke and Vilakazi give **unkungwini** as derived from *inkungu* (mist) and say it is a "species of *Plectranthus* plant used as a love-charm" (1958: 580).

And on that note, let us turn to love charms.

Love charms

One of the requirements for Hammond-Tooke's "persistence and well-being of society" is that humans meet, court, marry and procreate within the required cultural bounds. This goal, as we have seen above, is subject to malicious attack by various parties such as witches, and jealous co-wives and neighbours, and the use of plants to block such malice has already been described. But the process of courting may be derailed even before it starts if the quality of an

---

17. The first meaning is 'good snuff, good beer'; and 'good beer', in tempting the drinker to drink more and more, is surely also a cause of falling. It is likely that this plant name has a semantic link with *iwisa* (knobkerrie), given that this also causes great damage when it is 'caused to fall'!

aspirant suitor is not such as to attract the opposite sex. Bryant notes that the Zulu language has a number of words to describe such unfavoured men:

And there were bachelors, too, not indeed of choice, but of necessity; who, unblessed by the gods, lacked either the physique, the manner of life, the means or the luck to be attractive to females. Some lost their chance by being, as 'twas said, an *iDuma* (insipid); others, an *iHlule* (commonplace); others, again, an *iNyamfunyamfu* (dull, stupid), or an *amaQafukana* (coarse-featured), or an *isiKanavu* (clumsy, loutish), or an *amaNupungana* (dirty-looking), or an *iMfitimfiti* (ugly-faced), or a positive *iMpisi* (distinctly repulsive). But all alike, who were discarded by the girls, whatever was the cause, were derisively referred to as an *isiGwadi*, or *isiShinikezi*, or *isiShumanqa* (1967: 567).

For these unfortunate men there are a number of plants that can be used as love charms. Just one source – Pooley (1998) – lists 47 different species of plant used as love charms. Examples are **idunjana** or **ubukhoswane** (*Scadoxus multiflorus*), the **umkhokha**[18] (*Abrus precatorius*), the **umakhulefingqane** or **umdumbukane** (*Crassula vaginata*) and the **umkhokhothwane**, with the alternative names **umthongakazana**, **umsekelo** and **umnakile** (*Pyrenacantha scandens*). Of this last-named, Pooley says "Used in traditional medicine to prevent miscarriage, ensure an easy birth, treat impotence and barrenness, and as a love charm" (1998: 280). One sees in the use of this single plant all the necessary requirements for perpetuation of the species. The name **umsekelo** (support) could refer to any of its multiple uses, but the last name **umnakile** (having noticed him) is certainly a reference to its use as a love charm.

Many plants used as love charms are multifunctional in the same way. Consider the **impundisa** (also known as **intwalalubombo** or **umalibombo**, *Rubia cordifolia*) of which Pooley states "Used in traditional medicine as a mouthwash and to treat colic, sore throat, chest complaints, impotence, menstrual problems. Used by diviners to provide insight, protection. Love charm emetics" (1998: 558). The **iningizimu** (also **isidala** and **umzima**,

---

18. Pooley (1998: 60) says that **umuthi wenhlanhla** is an alternative name in Zulu and Xhosa for this plant, but as it means simply 'lucky medicine' I assume it to be a comment made by an informant about the plant rather than a name as such.

Dianthus zeyheri) is similar in its use as it is "used in traditional medicine for love charms, as charms to keep lightning away, and by diviners to improve their divining faculties" (Pooley 1998: 382). For those who wish to be attractive as lovers <u>and</u> enjoy 'all-round protection', the best bet is surely the **ibohlololo** (alternatively **uncolozi omncane** or **umjuluka**, *Aptenia cordifolia*) for Pooley says of this plant that it is "used in traditional medicine as a poultice, anti-inflammatory, deodorant, love charm and as protection against sorcery" (1998: 378). The name **umjuluka** (perspiration) is a direct reference to its use as a deodorant.

Hammond-Tooke refers to love charms in the following terms:

> Apart from the medicines used to ensure health, the problems of inter-sex relationships are catered for by an extensive corpus of love potions, philtres and charms to obtain and retain affection, which are widely used by all members of the community including members of the mission church in the district (1962: 275).

Bryant refers to plants used as love charms by the collective noun *ubulawu* and says that "of these, the Zulu youths possessed a whole pharmacopoeia, mostly emetic, and mainly herbal". He goes on to give the names of a number of such 'herbal remedies':

> So the amorous youth proceeded to 'wash himself'[19] internally with, for example, an infusion of *uMaguqu* roots, to make him 'feel nice'; others, of the *uQume*, to make him 'look lovable'; those of the *iPopomo*, to bring him 'good luck'; those of the *umNandi-wa-Vesha*, to ensure agreeable intercourse, and those of the *uNginakile*, to make her dream of him at nights (1967: 564).

These names are worth looking at in turn. The first – **umaguqu** – is derived from the ideophone *gúqu* (of turning over, changing). Doke and Vilakazi say of the noun **uguqu** that this is a "herbalist's name for the *isidenda* shrub *Maesa*

---

19. This idea of 'washing oneself' is well-reflected in the name *ungezakanye* that Doke and Vilakazi (1958: 555) give for a herbal love charm. The name is derived from the noun *ingeza* (purifying medicine, love charm to make one attractive) < *geza* (wash) + *kanye* (at one time).

alnifolia, the roots of which are used as a love-charm (lit. the changer)" (1958: 278). Bryant's name **umaguqu** is a morphological variant of the name **uguqu**.[20]

Doke and Vilakazi say of the plant **uqhume** that this refers to a "species of small bush, **Hippobromus alata*, whose roots are used as an emetic for young lovers and as a cure for headache" (1958: 704). They believe this name to be derived from the verb *qhuma* (burst, explode, pop) but to link it to either of the stated uses requires some imagination. Bryant gives the name *iPopomo* in his 1967 work (originally published in 1949), but it is not in his 1905 *Dictionary*. Doke and Vilakazi give **iphophoma** as "species of herb used as a love-charm" (1958: 670) and this is almost certainly Bryant's 'i-Popomo'. Doke and Vilakazi say that **iphophoma** is derived from the ideophone *phó*, but which of the three meanings they give they have in mind is anyone's guess: 1) of striking a knock-out blow, of hitting on the head; 2) of dropping inert, of falling dead; or 3) of letting fall.[21] The name **umNandi-wa-Vesha** is curious. The first part, 'umnandi', is clearly from the adjective *-mnandi* (pleasant, tasty), but 'Vesha' appears neither as a verb nor a noun in either Doke and Vilakazi or Bryant. If 'Vesha' is a personal name, the plant name would have the underlying meaning 'the tasty thing of Vesha'.

The last name is much clearer as *unginakile* simply means 'she has taken notice of me'. The name is similar to **umnakile**, which, as we saw a page or so back, is one of the names Pooley gives to *Pyrenacantha scandens*. Doke and Vilakazi have "**unginakile** (he is after me): species of forest climber *Pyrenacantha scandens* used to dispel nightmares, and to cause a girl to dream of her charmer" (1958: 555).

Many references to the use of plants as love charms suggest that they are used as 'love charm emetics' and we saw earlier that Bryant (1967: 564) refers to the young men's pharmacopoeia of love charms as "mostly emetic" and to the "amorous youth" who "proceeded to wash himself internally" with an infusion of *umaguqu* roots. There are, however, other ways of using plants as love charms. Hutchings explains how the plant **abanqonqosi** (*Podocarpus henkelii*) is used: "the bark is chewed and spat out into the wind as the name of the loved one is repeated" (1996: 14).

---

20. See 'Morphological variations' in Chapter 3.
21. Translating idioms and idiomatic language literally seldom works. What is an idiom in one language is not usually an idiom in another. But one can't help noticing, in this context, the English idioms and idiomatic language in 'she is a very <u>striking</u> girl', 'she <u>struck</u> me as very beautiful', '<u>falling</u> in love' and 'I <u>fell</u> for him right away'.

Pooley (1998: 50) notes of the plant *Hermbstaedtia odorata* (Zulu names: **ubuphuphu**, **umlwandle** and **uvelabahleke**) that when this plant is used as a love charm, it is mixed with the flesh of kingfishers. This possibly has something to do with the word *inhlunuyamanzi* (< *inhlunu* (vulva, vagina) + *yamanzi* (of the water)), the Zulu name of both the Malachite and the Pygmy Kingfishers (Doke and Vilakazi 1958: 339; MacLean 1985: 375, 376) a reference to the habit of kingfishers of plunging in and out of water.[22] The last Zulu name mentioned for this plant is **uvelabahleke** (< *vela* (appear) + *bahleke* (and they laugh, smile)), a name shared by a number of plants used inter alia for love charms, in reference to the effect these plants/love charms have on potential suitors. Doke and Vilakazi do not recognise the name 'ubuphuphu', but they do give *iphupho* (< *phupha* (to dream)) with the two meanings 'dream' and "any medicinal plant used by young men to make girls dream of them" (1958: 679). The plant name **umlwandle** is clearly derived from the noun *ulwandle* (ocean, sea), but it is not clear how this relates to the use of the plant as a love charm. It may well be a habitat name for this plant.

Returning briefly to the name **uvelabahleke** (he appears and they laugh), we note that this name is shared by a number of plants: the Wild Cockscomb (*Hermbstaedtia odorata*), the Heart-Leaved Brooms and Brushes (*Acalypha villicaulis), the Emerald Fern (*Asparagus densiflorus*), the Yellow Fire Lily (*Cyrtanthus breviflorus*) and the Hairy Lotononis (*Lotononis calycina*). Each of these plants has amongst its uses that of a love charm.

Love charms come with a variety of purposes in mind. A wife who thinks her husband is straying may purchase one to bring him back to the marital bed. A youth may doctor himself in order to enhance what he already thinks is great sex appeal. A girl may do the same, especially if she is past marriageable age and there have been as yet no suitors. But perhaps the most common intention of a love charm, if we are to believe the literature and the message conveyed by the names of these charms, is for a youth to attract the attention of a particular girl, and subsequently win and keep that girl. Such a charm is known as an *isibethelo*.

One of the more semantically transparent names for a plant used in love charms is **isiqomisa**, which Hutchings gives as a name for a species of

---

22. Cf. *ijekamanzi* (< *jeka* (have sexual intercourse) + *amanzi* (water)), the Zulu name for the dragonfly (Doke and Vilakazi 1958: 358).

Hybanthus saying that "root infusions are taken as love charm emetics by young men to give them a clean, healthy look" (1996: 205). Doke and Vilakazi give the meaning of *qomisa* as 'woo, court' and the derived noun *isiqomiso* as "medicinal charm used by young men to prosper [in] courting" (1958: 710). Equally clear is the name **ithandana** (love one another), which Pooley (1993: 110) records for the Natal Bush-Cherry (*Maerua nervosa*), although surprisingly enough, given the underlying meaning of the Zulu name, she makes no mention of the use of this plant in love charm medicines. From the verb *thanda* (love) comes the noun *intando*, which can mean both 'love' and 'love charm'.[23] To this *umuthi* (tree) can be compounded, giving **intandothi** (<*intando* + (*umu*)*thi*), a name that Pooley (1993: 318) gives for the Zulu Coshwood (*Cola greenwayi*). Again, Pooley makes no mention of the use of this tree in love charms.

Another name with an underlying meaning linked directly to love charms is **iphuphuma** (< *phuphuma* (be overcome with emotion)). Pooley gives this as a Zulu name for the Cape Holly (*Ilex mitis*) and this time she does note that it is "used as a love charm" (1993: 262).

A young man who has successfully courted and wooed is a happy young man and this is reflected in the name **ungcingci**, which Doke and Vilakazi give for a "species of flowering bush, used as a love charm" (1958: 551). The derivation, they say, is from the interjection *ngcingci* (How happy I am!) and they say the full name is **ungcingci-wafika-umntakwethu** (How happy I am that you have arrived, my sweetheart),[24] which is clearly what one says to a member of the opposite sex who has been washing with this flowering bush. This particular 'bush' would appear to be the 'trailing herb' *Polygala serpentaria*, for Pooley (1998: 404) records 'uncinci-wafika-umthakwethu' as one of three Zulu names for this species. The name **ungcingci** (i.e., without the arrival of the sweetheart) is not recorded by Pooley, Hutchings or Boon, but for the Natal Plane (*Ochna natalitia*) Boon (2010: 362) gives **umbovane-ngcingci** (and also **umbovu** and **umbovane**), while Hutchings (1996: 202) gives the name **umb(h)ovane-ongcinsi**.

---

23. Doke and Vilakazi say "love-charm used by women to secure the favour of their husbands" (1958: 784).
24. See Doke and Vilakazi (1958: 601) who give the meanings of *umntakwethu* as (1) member of our mother's hut, blood relation; (2) my sweetheart.

Ideophones are always used for dramatic effect in the Zulu language and this is certainly true of the name **ibhuzamlandela**, which Doke and Vilakazi give for "camphor, used as a love charm to cause girls to follow a young man" (1958: 56). The structure of the word is *bhúú* (of issuing together in numbers) + *za* (past tense subject concord for girls) + *m* (him) + *landela* (follow). The use of the ideophone *bhúú* creates a highly successful graphic image of a camphor-soaked young man being pursued by an ever-increasing horde of young girls. Doke and Vilakazi (1958: 887) also record this name as the more prosaic **uzamlandela**, which they translate as 'lo, they follow him', rather curiously turning the past tense concord for girls into the exclamation 'lo!'

We noted above, when charms used for protection against adverse weather were discussed, that Pooley gives **isiwisa** (make fall) as a name for *Centella glabrata* and says "used . . . as a charm against hail" (1998: 540). She gives the same Zulu name for the Spindlepod (*Cleome monophylla*), "used as a love charm emetic" (1998: 384). The same underlying meaning (i.e., cause to fall) is relevant and material from this plant is a main ingredient of a love charm that causes a girl to fall in love with you. Just as the Zulu name **isiwisa** echoes the English idioms 'falling for' and 'falling in love',[25] so the Zulu plant name **umalopha** (it bleeds) also finds an echo in the English expression 'my heart bleeds for you'. Pooley (1998: 164) gives this name for the Primrose Gentian (*Sebaea grandis*). The structure of this noun is *ma* (characteristically) + *l(i)* (it) + *opha* (bleeds). Hutchings notes that "whole plant infusions are taken as love charms emetics" (1996: 241).

The Large Orange Kalanchoe (*Kalanchoe paniculata*, Pooley 1998: 254) has the Zulu name **indabuluvalo** (that which tears up fear) with an underlying meaning that refers to the intended effect on the courting youth, who, tearing up his fear, essays out with the confidence that is the hallmark of a successful lover. Together with self-confidence goes self-determination, as expressed in the name **ungqangendlela** (what goes straight along the path; i.e., he who makes straight for the required objective). This is one of three names Pooley (1998: 404) gives for *Polygala serpentaria*. Another is simply **inhlanhla** (great luck), while a third is **ungcingci-wafika-umntakwethu**,[26] which, as we saw above, means 'How happy I am that you have arrived, my sweetheart!'

---

25. As mentioned in footnote 20 to this chapter.
26. I have corrected Pooley's spelling here.

If a courting youth wishes to catch the eye of a young lady, then what more suitable love potion to use than one in which an ingredient is the Starry Wild Jasmine (*Jasminum multipartitum*)? Pooley (1998: 164) gives as one of the Zulu names for this plant **ihlolenkosazane** with the underlying meaning 'eye of a young lady'. This same plant also has the Zulu names **imfohlafohlane** and **isandlasenkosikazi**, the former with the meaning 'a little breaking through' (getting through to the young lady, perhaps, or breaking through her reserves?), while the latter means 'hand of the lady', certainly a common target for any courting youth.

In our last two examples of plants used as love charms, namely *Polygala serpentaria* and *Jasminum multipartitum*, we noted that each plant had three Zulu names, each with an underlying meaning that referred to the use of the plant in love charm medicines. With some plants, the several names, when taken in combination, reveal an entire strategy of courtship. Take, for example, the Tree Orchid (*Cyrtorchis arcuata*) for which Pooley (1998: 130) gives the Zulu names **imfeyenkawu**, **iphamba** and **umbambela**. The first name has the underlying meaning 'sweet-reed of the monkey', suggesting that the love charm will make the suitor as attractive as, well, a piece of sweet-reed to a monkey. The noun **iphamba** is a simple derivation from the verb *phamba*, which means to 'trick' or 'confuse', perhaps 'bedazzle', the target. Should the youth be successful in his wiles, the young lady will become attached to the suitor, as suggested in the third name **umbambela** with its underlying meaning of 'grabbing for' and 'catching hold' of her.

Hutchings gives the same three Zulu names for this orchid and the relevance of the last-given (**umbambela**) is seen in her description of its use:

> used as an ingredient in a love charm emetic administered by a young man to his girl to make her affection return so that she will cling to him as the orchid clings to the tree (1996: 71).

This is interesting: presumably the name is given to the orchid because it clings fast to a tree. But is its use as a love charm (to make the girl cling to the young man) derived from the clinging behaviour of the plant or the underlying meaning of the name?

Umusa (graciousness, kindness, loving tenderness) is a name that Pooley (1998: 182) gives to the plant *Stachys nigricans* "used as an ingredient in love charm emetics" (Hutchings 1996: 267). This 'loving tenderness' may refer to the quality the young lover wishes in himself as a result of taking this potion, or it may refer to the qualities he is seeking in a young lady. The same applies to **umamatheka** (the smiling one), one of the names of *Lotononis corymbosa*: does this mean that the suitor wishes to improve on his own smile, or is he hoping that girls will smile on him?

For some of the more shy suitors, reaching for the young lady's hand, or hoping that she will fall in love with him, is an impossible dream. For such young men, simply getting the lady to unbend a little is a good start and for them the correct potion is one containing parts of the Forest Smilax (*Behnia reticulata*, Pooley 1998: 230) with its Zulu name **isigoba** derived from the verb *goba* (to bend over). For others again, all that is desired is a simple agreement between themselves and the chosen one, achieved perhaps by the use of *Sebaea sedoides* (no common name, Pooley 1998: 298) with the Zulu name **isivumelwano** (agreement).

There are plants that clearly must be taken at all stages of courtship, from the first meeting to the culmination of marriage. Let us look at *Eulophia cucullata* (no common name, Pooley 1998: 372), another plant with three Zulu names. The first suggests the initial stage of courtship: **uhlamvu lwabafazi** (lit., seed of woman) with its suggestion of planting the seeds of love. A successful stage of courtship is suggested in the second name **amabelejongosi** (the breasts of the young maiden), while the final stage is suggested in **udwendweni**, which simply means 'bridal party'.

There are some plants, clearly indicated in the literature as being used in the preparation of love charms, but which have names with an underlying meaning that is difficult, if not impossible, to relate to the whole business of courting. What does one make, for example, of the plant *Pelargonium luridum*, with its three names **inyonkulu** (a shortened form of **inyonenkulu** (big bird)), **isandlasonwabu** (the hand of a chameleon) and **unyawolwenkuku** (the foot of a chicken)?[27] And how do we interpret the Zulu names of another plant well known for its use in love charms – *Syncolostemon densiflorus* (the Pink Plume) – and known to Zulu *izinyanga* as **isidlekesenqomfi** (the nest of the Yellowthroated Longclaw) and **isolemamba** (the eye of the mamba)?

One needs to have a great deal of confidence in the *inyanga* when he is mixing up love potions with ingredients that may look like plant material, but

---

27. Hutchings says of this same plant, giving the Zulu names **inyonkulu, isandhla sonwabu, ishaqa** and **uvendle** that "Powdered dried roots, mixed with hippopotamus fat or python fat, are reported to be used by young men as courting charms, rubbed on the face" (1996: 149).

underneath are really the eyes of snakes, the hands of chameleons and the feet of chickens. For myself, however, I would be far happier to go out courting knowing that I have been fortified with a love charm containing bits of *Cyanotis speciosa*, also known as **umakotigoyile** (the hypnotised maiden).

Two extremely powerful love charm medicines are reflected in the underlying meanings of the names **ushisizwe** and **intombikayibhinci**. The first is recorded by Hutchings (1996: 95) for *Portulaca pilosa*, saying simply that it is "used as a love charm". The Zulu name means literally 'burn the country'.[28] The second name, **intombikayibhinci** (the girl does not wear clothes), is recorded by Boon (2010: 78) and Hutchings (1996: 75) for the Cape Fig (*Ficus sur*). My colleague Ndela Ntshangase says that "this love charm is so powerful that when the girl takes it and then sees you, she does not even wait to get dressed, or if she is dressed, she flings off all her clothes when she sees you".[29]

We end this section on love charms with two curiosities. The first concerns the following entry from Bryant's *Dictionary*:

> The fat of *umabhengwane* (Woodford's Owl) is mixed with *iSokalakwazulu* (common washing-soda) to make an *iHabiya* = (pg 220) medicine or love-charm of any kind (of modern introduction from Natal) used by young men to *hayiza*, i.e. to throw her into fits of shouting hysteria in which she repeatedly cries out *hayi! hayi!* or *hiya! hiya!* (1905: 371).

There are a number of queries here. How does 'common washing soda' come to have the quality of a love charm? Why should such washing soda get the name 'isokalakwaZulu' (the courting youth from KwaZulu)? Is there any significance in the choice of the *umabhengwane* owl as a source of fat to mix with the washing soda? And, finally, why would any young man seek to throw any girl into fits of shouting where "she cries out *hayi! hayi!* or *hiya! hiya!*"?

The second curiosity concerns the plant name **umnqologombotsheni**, which Doke and Vilakazi give for a "species of composite herb *Aster muricatus*, which grows between stone" (1958: 594). They also give the meaning 'effeminate man'. This is perhaps mildly curious. What is more

---

28. Returning again to the overlap between English idiomatic language and Zulu plant names, I am reminded of the rather dated English usage of the word 'flame' in such phrases as 'my old flame'.
29. Personal communication, March 2000.

curious is that **umnqologombotsheni** is a compound of two nouns – *umnqolo* and *ugombotsheni* – both of which mean 'effeminate man'. Does this mean an 'umnqologombotsheni' is twice as effeminate as most effeminate men? Is the *-tsheni* at the end of the word derived from Zulu *ematsheni* (among the stones), seeing that this is where *Aster muricatus* grows? Hutchings is unable to help us here as she does not include *Aster muricatus* in her book. Again, there are so many questions. The only thing one can be reasonably sure about is that an *umnqologombotsheni* is a person with little or no need for love charms.

Love charm emetics

In many of the descriptions above of plants used as love charms we have seen that quite a number have been described as 'love charm emetics'. For example, see the following nine entries from Pooley (1998):

- *Sebaea grandis*: Primrose Gentian, **umalopha omncane** "used as a love charm emetic" (164);
- *Stachys nigricans*: **umusa** "Used . . . in love charm emetics" (182);
- *Sebaea sedoides*: **isivumelwane esikhulu, umanqweyana, umsolo** "Used . . . as a love charm emetic" (298);
- *Utricularia prehensilis*: Large Yellow Bladderwort, **iphengulula** "used as a love charm emetic" (306);
- *Rubia cordifolia*: Sticky-Leaved Rubia, **impindisa, intwalalubombo, umalibombo** "Used by diviners to provide insight, protection. Love charm emetics" (558);
- *Gazania rigens*: Trailing Gazania, **ubendle** "used for a love charm emetic" (334);
- *Cleome monophylla*: Spindlepod, **isiwisa esiluhlaza** "used as a love charm emetic" (384);
- *Graderia scabra*: Wild Penstemon, Pink Ground Bells, **impundu, isimonyo, ugweja, umphuphutho** "Used . . . as love charm emetics" (430); and
- *Buchnera simplex*: False Verbena, **umusa omkhulu** "Used as a love charm emetic" (432).

Why should so many love charms be used in the form of emetics?

Theunis Botha lists no less than six small rivers with the name inHlambamasoka (1977: 180). It means 'where the courting youths wash' and the number of rivers so named clearly indicates that washing is very important before going out courting. In the river, however, one only washes the <u>outside</u> of the body. But as we noticed above, in the introduction to this section on love charms, Bryant pointed out that "the amorous youth proceeded to 'wash himself' internally with, for example, an infusion of *uMaguqu* roots, to make him 'feel nice'" (1967: 564). This, I think, helps to explain the importance of the 'love charm emetic'.

Plants used as charms for positive and practical purposes

> Dignity and the characteristic of fearfulness are acquired by men of authority through pounding heavy stones into a fine powder. Fat and the ground eyes of snakes which are feared are mixed with the powder and, if at all obtainable, quicksilver. The face and the limbs of the man are anointed with the mixture. He is believed to develop the required dignity and awesomeness. The symbolism lies with weight and its relationship to dignity. For heavy people (*abasindayo*) are prosperous and dignified (Berglund 1976: 355).

A number of plants are used as charms, not for purposes of protection against threats, but for more positive and practical purposes, such as making people respect you, making your business prosper, confusing an opponent in court, and so on. Many of these goals are reflected in the underlying meaning of the names of the plants so used. Let us look at some of these names. For assistance with the underlying meanings of the first five names, I am indebted to my colleague Ndela Ntshangase and our discussions about plant names.

Boon (2010: 548) gives **ukhukhulakhethelo** as a Zulu name for the Narrow-Leaf Butterspoon (*Coptosperma supra-axillare)*. Although this is not a name recognised by Doke and Vilakazi, they give *khukhula* as 'sweep away', 'carry before' and *ikhethelo* as 'the pick, the choice, the best' (1958: 410, 391). Ntshangase explains that when this plant is taken as a medicinal charm, nothing will prevent you getting exactly what you want.

Boon (2010: 444) gives 'umalungazalazikahkona' as one of the Zulu names of *Maesa lanceolata*. In correct Zulu this is **umalunguza-lazingakhona** (< *u* + *ma* + *lunguza* (peep) + *la*[*pha*] (there where) *zingakhona* (they are thereabouts) = peep in the direction where they are). Ntshangase says the plant is used as a charm to help you find what you are looking for.

The name **umdwendwelengcuba** is one of a number of names Boon (2010: 272) gives for *Searsia chirindensis* (Red Currant). This name is a compound of the nouns *umdwendwe* (line of people) and *ingcuba* (lean meat). Ntshangase explains this name: "when you use this plant as a medicinal charm, people will line up for your business". In this name, business is expressed metaphorically as 'meat'.

Boon (2010: 166) gives **umhlonishwa** as one of two names for the Narrow-Leaf Fountain-Bush (*Psoralea glabra*). Literally it means 'that which is respected' from the passive form of the Zulu verb *hlonipha* (respect). Ntshangase suggests that this plant, if used medicinally, will make people respect you. Hutchings, on the other hand, says the plant is "taken as emetics by healers afflicted with mental disturbances associated with their calling" (1996: 136).

The opposite effect is intended when medicines known generally as *isichitho* are used. These are medicines employed maliciously to make someone disliked by others. An example is **umsathanina**, a name given by Boon (2010: 376) for *Rawsonia lucida* (Forest Peach). As Doke and Vilakazi (1958: 724) point out, this word is a term of vulgar abuse. It is a compound of *satha* (have illicit sexual intercourse) and *unina* (his/her mother). Ntshangase says that this is used as an *isichitho* that will make people appear so ugly and evil that others will swear at them and insult them. Clearly used for the same purpose is *Pavetta gardeniifolia*, for which Boon (2010: 562) gives the two Zulu names **isinyombolo** and **isinywane**. Doke and Vilakazi (1958: 628) say that *isinyombolo* (unpopularity, disfavour) is another name for **isinywane**, which they give as "species of shrub, **Pavetta assimilis,* of the coffee family, used as a charm to bring a man into disfavour" (1958: 632). The English common name, Stink-Leaf Brides-Bush, is also suggestive in this context.

Doke and Vilakazi give the name **isalamuzi** (*sala* + *umuzi* (what stays in the kraal)) for a "species of herb, believed to be a means of procuring wealth" (1958: 721).[30] The underlying meaning is perhaps a little obscure when related to the use of the plant. Does it perhaps mean that when one has acquired wealth, one will keep it?

Also from the pages of Doke and Vilakazi (1958: 806) is the name **isithundu** (medicine calculated to bring prosperity). From this is derived the

---

30. This name is not to be mixed up with the noun *isalamusi* 'Malay person', derived from the word 'Islam'.

name *usithundu*, a love charm emetic. Hutchings gives **isithundu** as one of a number of Zulu names for *Cassine papillosa* and says "trees are believed to have powerful magical properties and unspecified parts are used to blunt the effects of evil spirits" (1996: 186). She also gives (1996: 202) **isithundu** as a name for *Ochna natalitia,* quoting Doke and Vilakazi about medicine used to bring prosperity.

Court cases can often be threatening not only to one's prosperity, but also to one's tranquility, and plant charms can be useful against such legal threats. Doke and Vilakazi say of the name **isindiyandiya** that it refers to "a species of coastal forest tree, *Bersama lucens* . . . taken as a charm to confuse one's opponents in court" (1958: 539). Boon (2010: 328) also gives the name **isindiyandiya** for *Bersama lucens* as well as the name **undiyaza**. According to Doke and Vilakazi, both names are derived from the ideophone *ndiya* (of being stunned, confused, giddy). They say that the name **undiyaza** refers to a "species of forest climber, *Dioscorea dregeana*, which causes madness" (1958: 539). No doubt this is equally useful in court, as long as it is one's opponents who are made to go mad.

Evil spirits and lightning can apparently also be confused and stunned by **isindiyandiya** for Hutchings (1996: 190) quotes Valley Trust healers as saying that preparations made from the bark of *Bersama lucens* can be used as a protective charm against them.

Another name that suggests the power of shutting up an opponent is **indodabindenye**, one of several Zulu names given by Hutchings for *Strophanthus speciosus* (Monkey Ropes). This is a compound noun with three elements: *indoda* (man) + *binda* (stifle, silence speech) + *enye* (another [man]). If your opponent should ideally be rendered ineffective in court cases, you yourself should aim for the opposite and ask your herbalist for **ingulamlomo** (what opens the mouth), a name given by Doke and Vilakazi for an unidentified medicinal plant "used as a charm to aid in law-cases" (1958: 564).

From prosperity and fluency in the courtroom we move to a more rural threat, that of snakebite. Many plant names containing the element *inyoka* (snake) have already been discussed in Chapter 7 on the medicinal use of plants. The discussion there revolved around the use of plants to treat people already bitten by snakes. The following five names refer to plants used in charms to keep snakes away before they have a chance to bite.

Pooley (1998: 508) gives the name **ihlinzanyoka** (skin a snake) to *Kniphofia parviflora* saying that this is used as a snake deterrent. Hutchings (1996: 31) gives the same information. Pooley (1993: 272) gives this name in the form 'uHlinzanyaka' for the tree *Putterlickia verrucosa*, but makes no mention of its use as a snake deterrent. Hutchings (1996: 183) also assigns the name **ihlinzanyoka** to various species of *Maytenus*, but also makes no reference to their use as snake deterrents. The name is used with the qualifying adjective 'small' as **ihlinzinyoka elincane** (little snake-skinner), a reference to the Small White Poker (*Kniphofia buchananii*) "used traditionally as a snake deterrent" (Pooley 1998: 92; Hutchings 1996: 31).

Doke and Vilakazi have **umnqandanyoka** (what keeps away a snake) for a "species of Senna weed *Cassia occidentalis* with a very repugnant smell" (1958: 590). They give **ivimbanyoka**, with the same underlying meaning, for the same plant. Hutchings gives *Cassia occidentalis* as an earlier synonym for *Senna occidentalis* and uses the Zulu name 'umnwandanyoka' saying that "leaves are used in snakebite" (1996: 130). Doke and Vilakazi recognise neither 'umnwandanyoka' nor a verb 'nwanda', so Hutchings' name here appears to be a spelling error.

Pooley has 'umqanda' (should be **umnqanda**) for *Kohautia amatymbica*, one of these multi-purpose charms; this one "Used in traditional medicine to improve appetite of infants, as a charm to protect babies from evil, as love charm emetics and protection against snakes, lightning" (1998: 202).

Pooley gives the name **ishaladi lezinyoka** (garlic of snakes) for Wild Garlic (*Tulbaghia acutiloba*), saying it is "used traditionally as a snake repellent" (1998: 510). Doke and Vilakazi (1958: 731) point out that **ishaladi** is an adoptive (loan word) from the English word 'shallot' and indicate that it has a number of distinct references to the world of vegetables. On its own it means 'onion', when suffixed with *elikhulu* (big), it means 'leek' and when suffixed with *lenyoka* (of the snake) it means 'garlic'. For Hutchings **ishaladi lezinyoka** refers to *Tulbaghia alliacea*, which is "cultivated to keep snakes away from the home" (1996: 37).

Hutchings (1996: 248) gives the name **umshayimamba** (strike the mamba) for *Strophanthus speciosus*. As this is used for snakebite, it has already been discussed in Chapter 7.

The Dune Poison-Bush (*Acokanthera oblongifolia*) is a plant with two names, each one reflecting one of its two purposes. Pooley (1993: 426) gives the names **inhlungunyembe** (poisonous slander) and **ubuhlungubenyoka** (snake's poison). The first name suggests protection against malicious gossip, while the second indicates protection against snakebite. Pooley says that "all parts of the plant contain a digitalis-like (heart) poison" and that it is "used medicinally against snakebite". Hutchings (1996: 242) says nothing about the use of *Acokanthera oblongifolia* as a charm against malicious gossip or slander, but she does note its use for snakebite, as well as for destroying marauding dogs.

Another 'oblongifolia' is the Dune Soap-Berry (*Deinbollia oblongifolia*) and Pooley (1993: 288) gives six Zulu names, of which three indicate their use as protective charms: **iphengulula, iqinisamasimu** and **umaqinisa**. The first name indicates general protection, most likely against evil: **iphengulula** (< *phengulula* (turn right over, 'strike back in defence'), while the other two tell of protecting the fields and crops: **iqinisamasimu** (what strengthens the fields) and **umaqinisa** (the strengthener)).

To protect the homestead generally, the gates can be metaphorically closed against evil intent and this is the basis of the name **umvalasangweni**. This name is used for three species of Gardenia: the Natal, the Transvaal and the White Gardenia (*Gardenia cornuta, G. volkensii* and *G. thunbergia*). Pooley says that the "Zulu name means 'to close the gate', [and the plant is] planted at homestead gates to keep evil spirits away and used to close cattle kraals" (1993: 460). In the form Pooley has given the name – 'umvalasangweni' – the meaning would be more 'what closes [something] in the gateway', a subtle but important distinction. Doke and Vilakazi give the name **umvalasangwana** (< *vala* (close) + the diminutive of *isango* (gate)) and say this is used of various species of small trees, "especially Gardenia thunbergia" (1958: 829). For Doke and Vilakazi, the purpose of the plant is not so much to use it to 'close the gate against evil', but actually to <u>make</u> garden gates.[31]

Use of plants in rites and rituals

Many are the various uses of plants for rites and rituals, but few are the actual names with an underlying meaning referring to such uses.

Doke and Vilakazi have **isithushana** (< diminutive of *isithutha*; i.e., *idlozi*) for a "species of tree, *Weihea gerrardi*" (1958: 809). With the same underlying meaning is **umadlozane**, which Boon (2010: 250) gives for *Margaritaria discoidea*. This is **umadlozini** in Hutchings (1996: 63) from the locative form *emadlozini* (among the ancestors).

Boon (2010: 324) gives **umuthi-wezithutha** (medicine of spirits/ancestors) as a name for *Deinbollia oblongifolia*. Hutchings also gives the name **umuthi**

---

31. See the sub-heading 'Enclosures' in Chapter 9.

wokuzila[32] (medicine of avoidance) for this same tree and says "strengthening medicines from the plant are taken after the death of a kraal member" (1996: 189). The Zulu verb *zila* has three related meanings: 1) fast, abstain, avoid; 2) show respect to; 3) mourn, observe mourning ceremony.

Pooley has **umsuzwane** as a Zulu name for *Lippia javanica* and says "used for ritual cleansing after contact with a corpse and for protection against dogs, crocodiles, lightning" (1998: 180). Doke and Vilakazi give the name as **umsuzwane** (< passive *suza* (fart)) and say "species of shrub, *Lippia asperifolia*, having a very disagreeable smell, used for smearing the body as a protection against crocodiles and dogs" (1958: 771).

Although the name **usingalwesalukazi** means 'the thread of the old woman' and is discussed in Chapter 9 under the sub-heading 'Fibres', it is worth mentioning that Hutchings says of this plant (*Asclepias physocarpa*) that "seeds are blown away from the pods as a charm to placate the ancestors" (1996: 253).

The Buffalo-Thorn (*Ziziphus mucronata*) has the Zulu names **umphafa**, **umlahlankosi**, **umlahlabantu** and **isilahla**, which is an abbreviation of the last two. **Umlahlankosi** and **umlahlabantu** can be translated as 'what buries the chief' and 'what buries people' respectively. Some sources say this tree is planted above the grave of a chief; it is perhaps better known for its role in the *ukubuyisa* (bring back) ceremony, held a year or so after death, to bring back the wandering spirit of the deceased to the homestead where the deceased lived. In this ceremony, a branch of the tree is dragged around the circumference of the homestead while the new head of the homestead calls out the clan praises and personal praises of the deceased, begging the spirit to return home (see Plate 22).

---

32. This may not actually be a name, but an explanation by an informant as noted in footnote 17 for *umuthi wenhlanhla* (lucky medicine).

9 Names referring to the practical use of plants

Food plant of . . . emperor moth caterpillars [which] are collected, roasted and eaten (*amacimbi*) . . . trees are retained when clearing fields because they provide food & shade. The abundant crops of juicy, tart fruit, high in vitamin C, are eaten by people and animals. Used to brew a refreshing & intoxicating drink (*ubuganu*) . . . A jelly preserve is made from juice. Amarula liqueur is produced commercially. Woody stones are laboriously cracked to get nut-like seeds, which are small, highly nutritious & tasty . . . seeds are stored, eaten raw or cooked with maize meal . . . Bark is used medicinally & provides a light brown dye used in basketware. Wood pinkish-white, used for carvings, fuel & furniture (Boon 2010: 268).

The quotation above refers to the Marula (*Sclerocarya birrea* subs. *caffra*) or **umganu** in Zulu. Not every tree has quite the same wide range of practical uses as the Marula, but several come very close. This chapter looks at such practical use of plants in Zulu society and then goes on to discuss the Zulu names of plants that reflect such use.

The practical uses of plants

Besides their medicinal use and use as protective and other charms, as described in Chapters 7 and 8, plants are used for a wide variety of practical household purposes. Van Wyk and Gericke (2000) give a good idea of this variety, dividing their book into the three main sections 'Food and drinks', 'Health and beauty' and 'Skills and crafts'. The first section is divided into six chapters: (1) Cereals, (2) Seeds and nuts, (3) Fruits and berries, (4) Vegetables, (5) Roots, bulbs and tubers and (6) Beverages. The section 'Health and beauty' is mainly concerned with the medicinal use of plants, but there is a chapter on 'mood and mind plants' and two chapters deal with perfumes, repellents, soaps and cosmetics. Further chapters deal with (7) General medicines, (8) Tonic plants, (9) Mind and mood plants, (10) Women's health, (11) Wounds, burns and skin conditions, (12) Dental care, (13) Perfumes and repellents and (14) Soaps and cosmetics. The section 'Skills and crafts' covers the following: (15) Hunting and fishing, a chapter devoted almost exclusively to the use of poisonous plants, (16) Dyes and tans, (17) Utility timbers, (18) Fire making and firewood, (19) Basketry, weaving and ropes and (20) Thatching, mats and brooms.

Given this wide variety and very extensive use of plants in the daily life of humans, one would expect there to be a corresponding range of plant names with underlying meanings referring to such usage. But in fact, compared to the number of names with underlying meanings referring to the plant itself, and to those referring to medicinal use and use of plants as love or protective charms, there are few names that refer to the use of plants as food and drink, as timber and firewood, and as sources for weaving and thatching (to mention just a few of the uses given by Van Wyk and Gericke).

Categorising and classifying plants according to their use in society is difficult, for at least two reasons. First, it is difficult to balance a categorisation based on the plants, or the parts of the plants actually used, with the human need satisfied by the plant. See, for example, in Van Wyk and Gericke's Section 1, where 'Drink' is a human need category, but 'Cereals', 'Seeds and nuts' and 'Fruit and berries' are categories of plant type and plant parts. Second, as we have seen from the quotation on the usage of the Marula Tree at the beginning of this chapter, one particular species may have several different uses and so appear under many different headings. In Van Wyk and Gericke's categorisation, the Marula would appear in Chapter 2 'Seeds and nuts' ("nut-like seeds, which are small, highly nutritious"); in Chapter 3 'Fruits and berries' ("abundant crops of juicy, tart fruit"); in Chapter 6 'Beverages' ("refreshing & intoxicating drink"); Chapter 7 'General medicines' ("bark is used medicinally"); Chapter 16 'Dyes and tans' ("bark . . . provides a light-brown dye"); Chapter 17 'Utility timbers' ("Wood . . . used for carvings . . . furniture"); and Chapter 18 'Fire-making and firewood' ("Wood . . . used for . . . fuel"). This must not be seen as a criticism of the categorisation system used by Van Wyk and Gericke (2000), but simply as an example of the potential difficulties in balancing human needs, plant types, plant parts and the multifaceted use of so many plant species.

There is, however, some connection between plant part and the way humans utilise the plant, particularly in the case of trees. Leaving out medicinal use and the use of tree parts for love charms and protective charms, the different parts of a tree may be used for the following practical purposes:

- The wood (timber) is used for firewood, construction, fence poles, carving and the manufacture of a variety of implements;
- The fruit is eaten and used to make preserves, jams, jellies and both alcoholic and non-alcoholic drinks;
- The seeds are eaten raw, roasted, ground into flour, and provide oils;
- The sap can occasionally be tapped to make wine, but usually provides glue, bird lime, fish poisons and insecticides;
- The bark usually provides twine and cordage, dyes and is used for tanning;
- The leaves may be eaten and are also often used cosmetically;
- Smaller twigs and branchlets have a number of uses, from the construction of fishing baskets to their use as toothbrushes; and
- Roots are occasionally ground into flour and eaten.

Let us now turn to the actual names of trees and other plants that refer to the different kinds of practical use mentioned above.

Names referring to practical uses

Uses of wood

Many Zulu tree names that refer to their use as timber, whether as firewood, for construction, enclosures or other purposes, have already been mentioned in Chapter 6 under the sub-heading 'Tough, hard wood'. Here is a brief summary of them:

- **umhlulambazo** ('what overcomes the axe'): the Natal Box (*Buxus natalensis*);
- **umzilazembe** ('what shuns the axe'): the Sickle Bush (*Dichrostachys cinerea*); **umdlulamazembe** ('surpasses the axes'): the Common Pheasant-Berry (*Margaritaria discoidea*);
- **umhlabambazo** ('what stabs the axe'): the Dune Bride's Bush (*Pavetta revoluta)*;

- **umazwenda** (< *uzwenda* (tough object)): the Red Hook-Berry (*Artabotrys monteiroae)*;
- **umuthinzima** (hard tree): the False Soap-Berry (*Pancovia golungensis*); and
- **umanzimane** (< *nzima* (tough)): the Natal Worm Bush (*Cadaba natalensis*).

Many of these tree names use metaphors, as in:
- **ithambo** (bone): the Galla Plum (*Haplocoelum foliolosum*) and the Forest Elder (*Nuxia floribunda*);
- **ithambolempisi** (hyena's bone): the Common Canary-Berry (*Suregada Africana*);
- **impondondlovu** (elephant's tusk): Leadwood (*Combretum imberbe*);
- **umsimbithi** (< **(in)simbi** (iron) + **(umu)thi** (tree)): the Umzimbeet (*Millettia grandis*); and
- **umthombothi** (*Spirostachys africana*) with Van Warmelo's theory that *umthombo* is an obsolete Zulu word for 'stone', giving us 'the tree with wood hard as stone'.[1]

Firewood

Perhaps the most important economic function of wood, and certainly the most common daily usage, is the use of wood for the twin purposes of cooking and keeping warm. Not all wood is used as firewood: some tree species (e.g., **umthombothi**, *Spirostachys africana*) contain toxic oils and resins and it can be dangerous to cook with this wood or to inhale the smoke. Other trees are avoided for spiritual reasons. Some trees burn well even when green, some burn very quickly, others very slowly and some are best used as charcoal. Some species are especially useful when making fire by friction and others as fire starters or kindling. Boon specifically mentions 61 species regarded as useful firewood. Van Wyk and Gericke have the following to say about firewood:

> Firewoods are typically heavy and hard, and differ from utility timbers in their high energy value, their ability to form coals rather than ash, and their slow burnout time. They should also be free from poisonous substances (2000: 283).

---

1. See page 65.

An obvious name referring to firewood is **umthezane** (< *theza* (collect firewood)), a name Pooley gives to the Sickle Bush (*Dichrostachys cinerea*) stating "wood excellent for . . . firewood" (1993: 142). Coates Palgrave enlarges on this with "It is an excellent firewood, burning well but not too rapidly" (1977: 254).

For the False Currant (*Allophylus decipiens*), Boon (2010: 320) gives the Zulu name **umcandathambo** (< *canda* (chop up firewood) + *ithambo* (bone)). Clearly this is a wood that will burn well and last, for Coates Palgrave (1977: 526) says the wood is hard, white and dense.

With the availability of matches even in remote rural areas of KwaZulu-Natal, it is doubtful if there is still a need to make fire by friction, but from the evidence of plant names it is clear that this was once a regular practice. Doke and Vilakazi (1958: 652) explain the name **uphehlacwathi** with the underlying meaning 'what spins the firestick'.[2] Boon (2010: 514) gives this as a name for *Clerodendrum glabrum, with the equally relevant English vernacular name Tinderwood. Coates Palgrave does not mention the use of wood from this tree as being used to start a fire. In fact, he rather puts the fire out when he says "this is one of the so-called 'rain trees', a phenomenon caused by a small bug" (1977: 815).

Also referring to the starting of fire by friction is the Zulu name **uzwathi** (fire stick).[3] Boon (2010: 534) records this Zulu name for *Mackaya bella* (Forest Bell-Bush, River Bells) and also gives the name **umavuthwa** (the passive form of the verb *vutha* (to flame; i.e., that which is made to flame). Coates Palgrave says of the tree that "the wood is used to kindle fire by friction" (1977: 837).

When the fire has died down, the ashes remain. The underlying meaning of the Zulu tree name **isibangamlotha** is 'that which causes ashes' and this name is one of four Boon (2010: 230) assigns to the Tassel-Berry (*Antidesma venosum*), saying that the wood is used for firewood.

A slightly different function in the setting alight of wood can be seen in the name **ubhaqa** (lamp, torch, light, candle) one of the names Boon gives for *Ptaeroxylon obliquum* (Sneezewood). The wood of this tree is "very inflammable, [with] paraffin-like qualities even in green wood" (2010: 328).

---

2. Van Wyk and Gericke (2000: 290, 291) have interesting information and pictures on 'spinning a firestick' to start a fire.
3. Doke and Vilakazi (1958) give both *ucwathi* and *uzwathi* as words for a firestick.

Thorny firewood features in a Zulu proverb: the Kei-Apple (*Dovyalis caffra*) has the Zulu name **umqokolo**, found in *Woze utheze umqokolo, ukholwe ngameva* (in the end you'll gather thorny firewood, and know it by the thorns; i.e., You will learn by bitter experience) (Doke and Vilakazi 1958: 709).

Construction

Despite the fact that several dozen species of trees are mentioned in the literature as used for construction of various sorts, hut building and furniture making being the most common, there are very few names that refer to this.

The name **usahlulamanye** (what overcomes the others) refers to the hard wood of *Kiggelaria africana* (Wild Peach), which is used for furniture, floorboards and rafters; while **umkhandangoma** (make a drum) is more specific in its reference to "highly prized . . . for furniture" (Pooley 1993: 46) of *Podocarpus falcatus* (Yellowwood).

I am making a guess when I say the name **umzithi** is derived from *umuzi* (homestead) + *umuthi* (tree). The name is given to *Cleistanthus schlecteri* (False Tamboti) said by Boon (2010: 232) to be used for hut building.

Implements, utensils, craftwork

At first glance, the Zulu name **umvuma**, derived from the verb *vuma* (to agree), appears to have little or no significance as a plant name under the sub-heading 'implements, utensils and craft-work'. Boon (2010: 54) gives this as a Zulu name for *Raphia australis* (Raphia Palm).[4] Boon notes that the leaves of this palm are used for hut building, roofing and walls, and the buoyant leaf stalks are used for rafts. What is significant is that the Zulu verb *vuma*, according to Doke and Vilakazi, is an old word derived from the Ur-Bantu verb *lûma* (to allow) further interpreted by Doke and Vilakazi as meaning "turn out well (as from a mould, baking, brick-making, pottery, cooking, hide-dressing, etc.)" (1958: 841). The name is thus an old link between plant and human creative activity. Note that Boon (2010: 326) also gives this name to the Jacket-Plum (*Pappea capensis*). The detail given by Coates Palgrave (1977: 535) shows that this tree has a wide range of practical uses, ranging from edible fruits to the considerable number of uses of the oils derived from the seeds. So as with

---

4. Coates Palgrave (1977: 70) has Kosi Palm.

the Raphia Palm above, the Zulu name **umvuma**, with its historical Ur-Bantu root *lûma*, links the tree to its usefulness to society.

In similar fashion, the Zulu name for *Hyphaene coriacea*, **ilala**, has ramifications that extend into history and link plant and plant name to ethnic identity, blacksmithing and general craft skills.

Wright (2012: 361, 362) lists six interpretations of the name iLala (plural amaLala) as an ethnic term used in Natal and Zululand in the early years of the nineteenth century:

- people who lived mainly to the south of the uThukela River, particularly in the coastal region, as far south as the uMkhomazi River;
- people who spoke one or other variant of the *tekeza* dialect, in contrast to the *zunda*-style language of the Zulu heartland north of the uThukela River;
- people who smelted and forged metal;
- some 30 or 40 'tribes' (i.e., clans – people using the same *isibongo* as a surname) identified by various sources in *The James Stuart Archive* (Webb and Wright 1976–2014);
- an insulting term designating low social status; and
- an insulting term designating *tekeza*-speakers, this use coming into being only during the time of Shaka.

Of all these meanings, the one that interests us in the context of Zulu plant names is the third. Bryant (1905: 346) still gives the meaning '[black]smith' for *ilala*, but Doke and Vilakazi (1958) do not, suggesting that this meaning of the word had fallen out of use early in the twentieth century. Wright, though, puts forward Hedges' suggestion that "the term *ilala* had originally denoted a skilled craftsman and not a member of a particular [ethnic] group" (2012: 366). If this is so, then the word for this particular tree – *Hyphaene coriacea* – has links to a far broader range of craftwork than simply making baskets out of Lala Palm leaves, as seems to be the case with the name **umvuma** for the Raphia Palm and the Jacket-Plum (see Plates 7 and 8).

Another set of words with links to craft skills, plants and an early ethnonym is that based on the verb *thunga* (sew or stitch, with connotations of thatching and general weaving). Doke and Vilakazi (1958: 806) say this is derived from the Ur-Bantu verb root *-tuŋga* (pass through, a meaning that could cover a number of skills like sewing, thatching, and basket and mat making). From the similar Ur-Bantu noun *tuŋga* (basket) comes the modern Zulu noun, *ithunga* (wooden milk pail), showing a semantic shift from one kind of domestic container to another. Derived from the Zulu verb *thunga* is the noun **ithunga**, a species of grass (unidentified by Doke and Vilakazi) used for both thatching and basket making. In the passive form, the verb *thunga* is *thungwa* and this root appears in both the ethnonym iNtungwa (plural amaNtungwa) and the variant form **intungwa** for the thatching and basket making grass mentioned above. Although neither Bryant (1905) nor Doke and Vilakazi (1958) mentions **intungwa** as a grass or iNtungwa as an ethnonym in their respective dictionaries, all six volumes of *The James Stuart Archive* (Webb and Wright 1976–2014) make continual references to amaNtungwa. The relationship between this name for the ruling, elite group of Zululand at the time of Shaka and a type of grass is explained by one of Stuart's sources talking about the use of *insinde* and *intungwa* grasses for weaving and thatching:

> I have asked amaNtungwa people the origin of their name, and they said *it originates from the intungwa grass* (*entungweni yotshani*).[5] This grass will stick in clothes and prick one. That is, the name arose from the grass used for thatching huts. *Grain-baskets* (*izilulu*) are also made of *intungwa* grass (Webb and Wright volume 4, 1986: 176).[6]

But to get back to the Zulu plant names that refer to the usefulness of trees in craftwork, the Pride-of-De-Kaap (*Bauhinia galpinii*) is according to Pooley (1993: 150) used in basket making. She gives it the Zulu name **usololo** (pliant, unbreakable object).

Two plants used for making brooms have this reflected in the name *Gnidia caffra* (no common name) (Pooley 1998: 292) or **umshanyelo** (broom); and the Bushman's Tea (*Athrixia phylicoides*) (Pooley 1998: 442) with the similar name **ishanelo/ishayelo** (broom).

No doubt any small leafy branchlet may be used to whisk away flies and other annoying insects, but it is the Sand Peawood (*Craibia zimmermannii*) that has this function specifically referred to in its Zulu name **umphungwane** (little fly-whisk < *phunga* (flick away flies)). Van Warmelo (1976: 93) gives the similarly based Venda name *bunga-nyunyu*, the name of a small plant. He

---

5. Webb and Wright use italics when Stuart has recorded something in his original notebooks in Zulu and the editors have translated it into English.
6. This is also highly significant. Source after source in the six volumes of *The James Stuart Archive* says of the amaNtungwa people that *behla ngesilulu* (they came down with the grain-basket).

says the name is formed from the verb *fhunga* (fan away flies)[7] and the noun *vhu-nyunyu* (gnats).

Among many other Zulu names, nine in all, Hutchings (1996: 238) gives **umphatha-wenkosi** and **umphathankosi**, as well as **umphathankosi-omhlophe**, for *Strychnos decussata* (Cape Teak). The names are derived from the verb *phatha* (carry in the hand) and *inkosi* (chief) and mean the same thing – 'that which is carried by the chief'. Coates Palgrave says "in the 19th century [these trees] provided Zulu kings with their sticks of ceremony and gave the plant its local name of 'king's tree'" (1977: 765). In addition to the English name Cape Teak, Hutchings records Chaka's Wood and Panda's Walking Stick Tree.[8] The name **umkhangala**, recorded for this tree by Pooley (1993), Boon (2010) and Hutchings (1996), is found in the Zulu saying *Uyokuncinda uthi lomkhangala* (you will get a taste of a teak-stick; i.e., You will meet with misfortune) (Doke and Vilakazi 1958: 379).

Three Zulu names are given for *Oncoba spinosa*, the Snuff-Box Tree.[9] The name **umthongwane** simply identifies the species, but the names **umshungu** and **isingongongo** (< *ngóngongo* (of rattling)) are relevant to the uses of the hard shells of the fruit. These shells are used for rattles for dancers, snuff boxes and protective penis covers (see Plate 24).[10] The Zulu word *ishungu* (snuff box) is linked to the tree name **umshungu**, and not the other way around according to Doke and Vilakazi, who say of **umshungu** "Species of tree from which snuff boxes are procured" (1958: 748).

The importance of crops from the field and the smooth running of agriculture is reflected not just in names like **iqinisamasimu** and **umaqinisa** (described in Chapter 8) but also in the name **ugejelibomvu** (red hoe), given by Boon for the Assegai Tree (*Curtisia dentata*), of which he says "wood tough, hard, heavy . . . <u>dull red</u> used in the past for furniture, rafters and flooring" (2010: 438). Coates Palgrave (1977: 710) links this tree to its Zulu name in his words "wood hard, strong, elastic tough . . . suitable for tool handles". Another Zulu name linking tree and (agricultural) implements is **isandulana**,

---

7. Cognate with Zulu *phunga*.
8. 'Panda' was the form often used by early explorers and traders in Zululand for King Mpande, half-brother of King Shaka. See Koopman (2013b).
9. Coates Palgrave (1977: 624) gives the unlikely 'Rhodesian' name for this tree as 'Fried-Egg Flower'! Beaumont (personal communication) explains that the flowers look like fried eggs as they have large white petals with lots of yellow stamens in the middle.
10. For further information on the use of plant material as penis covers in Zulu society, see Koopman (2013c).

a name that Doke and Vilakazi (1958: 10) say is derived from the verb *andula* (commence cultivation).[11] The name **isandulana** refers to the Turkey-Berry Tree (**Canthium mundianum*), which Boon describes as having "wood hard, fine-grained . . . used for . . . tools" (2010: 542) and Coates Palgrave says "makes good fencing posts, furniture and implements" (1977: 884).

Enclosures

The name **umvalasangweni** that Boon gives for *Gardenia cornuta* (Tonga Gardenia) is an interesting one. Besides being used as firewood, the tree also provides poles used for fencing and Boon says the "Zulu name *umvalasangweni* means 'close the gate'. Planted at homestead gates to keep evil spirits away and used to close cattle kraals" (2010: 548). This is clearly a bi-lexemic word, but not in the sense of the polysemic words described in Chapter 4 where a word may have two or more different meanings related in one way or another. This tree name has one meaning – 'that which closes the gate'[12] – but it operates on two levels. On the physical level, branches or poles from the tree may be used as part of an actual gate; on the metaphysical level, the tree may be planted near to, or on either side of, a gateway, where its 'magical' powers will serve to block the entry of malevolent spirits.

As Pooley (1998: 158) explains, the scrambling, tree-smothering shrub *Pereskia aculeata* (Barbados Gooseberry) is an invader from Central and South America and would not have had a Zulu name when it first appeared in KwaZulu-Natal. However, as it now has the function of providing security fencing and protection for graves, it needs a Zulu name. The name **ufenisi** is an adaptation of the English 'fence'.

Trees providing food and drink

Fruit

Pooley (1998: 140) gives **umgandaganda** (pounding, mashing) for the scrambling shrub *Albertisia delagoensis* (no common name) and says of the velvety orange fruits that they are soaked in water and mashed before being eaten.

---

11. This verb is also the base of the second month *uMandulo* in the Zulu lunar calendar (Koopman 2002: 251).
12. Or strictly speaking, as pointed out in chapter 8: 'that which closes up in the gateway'.

A very simple name for an edible fruit is **umgwinya** (< *gwinya* (to swallow)).[13] This is the name Boon (2010: 446) gives for the Fluted Milkwood (*Chrysophyllum viridifolium*). Coates Palgrave says the fruits are "edible, with a pleasant taste" (1977: 726).

Apart from just swallowing fruit, one can also crunch them noisily, as in the name **ingqumuza** (< *gqumuza* (crunch food noisily)) for the Red Thorn-Pear (*Scolopia mundii*), of which Boon says "Fruit eaten by people" (2010: 378).

Both the plucking and eating of fruit are reflected in the name **ukhamginqi**, which Doke and Vilakazi (1958: 378) translate as 'pluck and devour'. This name refers to the Sourberry Kei-Apple (*Dovyalis rhamnoides*), which has "tasty fruit eaten by people . . . used to make jelly preserve" (Boon 2010: 372). Coates Palgrave (1958: 640) says the fruits are pleasantly acid-flavoured and much sought after.

The following two examples have already been listed in Chapter 6 and I repeat them here in much abbreviated form:

- **isibinda** (get stuck in the throat): the astringent, mouth-drying Cat-Thorn (*Scutia myrtina*, Afrikaans *droog-my-keel*); and
- **uguguvama** (precious thing found abundantly): *Lantana rugosa* with its juicy purple fruit clustered in bracts, eaten by people, monkeys and birds (Pooley 1998: 422).

Other foods

Hutchings (1996: 300) notes of the *Pavetta* species known as Christmas Bush, with the Zulu name **isamunyane** (bitter thing), that its leaves are chewed as a bitter tonic.

Another name referring to the eating of leaves is **u(lu)nquntane** (< *nqunta* (nip off, pluck off the tops), which Doke and Vilakazi give for a "species of climbing plant, *Riocreuxia torulosa* & *R. polyantha*, whose leaves are eaten as greens" (1958: 506). Pooley gives this name as one of three Zulu names for *Riocreuxia torulosa* and says that the plant is "eaten as a potherb, but the 'spinach' or *imifino* from these leaves is sometimes reputed to cause headaches" (1998: 176).

---

13. This verb is also the base of the noun *igwinya* (plural *amagwinya*), a popular oil-fried doughnut known as *vetkoek* (fat cake) in Afrikaans.

Boon (2010: 382) gives **isishwashwa** as a Zulu name for *Xylotheca kraussiana* (the African Dogrose) and says that people suck the arils off the seeds. One wonders if this has anything to do with the ideophone *shwá* (of spurting).

The Swazi Ordeal Tree (*Erythrophleum lasianthum*) (Pooley 1993: 144) has the Zulu name **umbhemise** (make him smoke). The verb *bhema* literally means 'take tobacco', whether in the more modern sense to smoke a cigarette or, in the older usage, to take snuff. *Bhemisa* is the causative form of the verb, meaning 'get someone to *bhema*'. Since Doke and Vilakazi under the entry for **umkhwangu** say of this tree that the "poisonous pungent bark is used as snuff for headache" (1958: 421), it may be more correct to translate the word **umbhemise** as 'give him snuff'. Doke and Vilakazi also note that the bark may be used for countering witchcraft.

Practical household uses

Dyes and tanning

Boon (2010) mentions only one tree as a provider of ink (*Sclerocroton integerrimus*), but the Zulu names given for this tree[14] make no reference to this. Pooley (1998: 52) gives the Zulu name **umnyandla** (< *unyandla* (secret message or messenger)) for the equally relevantly named Inkberry (*Phytolacca octandra*), saying that the ripe fruit was used in the past as ink.

It would seem from the name **umphendulobomvu** (turning red) that the Grey-Leaved Indigo (*Indigofera velutina*) (Pooley 1998: 390) is used as a dye, but there is no record of this in the literature. It may be that the name refers to the flowers, which according to Pooley are pink to rose red. Van Wyk and Gericke say of indigo that it is "one of the oldest colouring agents known to man, and that the leaves and roots of the African Indigo (*Indigofera arrecta*) are a source of indigo, which is used for dyeing" (2000: 254). They give the Zulu name of this tree as **umphekambedu**. Doke and Vilakazi recognise this name for the tree and give as an underlying meaning "what cooks pericardium fat" (1958: 652), which at first glance seems to have nothing whatsoever to do with the use of indigo as a dyestuff. But although the verb *pheka* means 'to cook', it also has the meaning 'to mark or dye with indigo colour' and although the

---

14. **umdlampunzi** (what the duiker eats) and **umhlepha** (no underlying meaning).

primary meaning of *izimbedu* is 'fat attached to the pericardium', the word is also used to indicate something contested for (as the fat is by boys), especially in the phrase *ubhedu olubangwayo* (a costly precious thing). So the name of this tree could be 're-translated', this time with an underlying meaning more like 'dyeing some precious thing with indigo' or 'a costly, precious thing that is dyed with indigo'.

Boon (2010: 196) says that the rootstock of *Elephantorrhiza elephantina* (Dwarf Elephant-Root) is used to tan leather. Van Wyk and Gericke who, incidentally, give the same plant the English name Elandsbean, go a little further:

> The underground rhizomes, commonly referred to as roots, are dug up
> and are used in rural areas for tanning . . .
> To dye grass for weaving, the pounded rhizomes are boiled with the
> grass for several hours, giving a brown or reddish colour (2000: 250).

The Zulu name for this plant – **intolwane** – in itself contains no reference to the use of the plant as a red dye, but its use in a well-known praise phrase for a red bull does. As I have explained elsewhere,[15] the name **uJamludi**, derived from *Jan Bloed*, a common early Dutch name for a red ox, is often found in the phrase *uJamludi obomvu njengentolwane* (Jamludi who is red like the intolwane shrub).

Fibres

I can find only three names that refer to this specific use of plants, and the third name is perhaps a little suspect.

Pooley (1998: 168, 546) gives **usingalwesalukazi** as a Zulu name for *Gomphocarpus physocarpus* (Milkweed, Balloon Cottonbush, Hairy Balls) and says the "stems [are] used for fibre (Zulu name refers to the fibre used for sewing *isidwaba*, leather skirts of old women)".[16] Hutchings (1996: 253) does not mention the sewing of *izidwaba*, but does say that stripped green bark is tied around the waist of new-born infants with apparent urinary problems and adds the English name Bindweed to the three English names given by Pooley.

---

15. See Koopman 2002: 215-216 and 2005.
16. 'isidwaba' should be the plural form *izidwaba*. These leather skirts are worn by married women rather than 'old women'.

Pooley (1998: 280) also gives **usingalwesalukazi** for *Euphorbia kraussiana*, but makes no mention of the use of this plant for fibre.

For the Brown Gonna (*Passerina filiformis*), Pooley has the name **unwele oluncane** (small single hair). She notes that this plant has "tough fibrous bark used for binding twine" (1993: 342).[17]

Boon (2010: 84) says of the Mountain Nettle (*Obetia tenax*) that the bark produces strong fibre. It is not clear if this is reflected in the Zulu name **impongozembe**. Doke and Vilakazi (1958: 669) give the meaning of *impongoza* as 'long, limp, rope-like mass' and this could be suffixed by the ideophone *mbé* (of sticking fast, firm) or the adjective *mbe* (another, different one).

Cosmetics and personal hygiene

Isimonyo (< *monya* (smear the face)) is a Zulu name Pooley (1998: 290) gives for the Small Hypericum (*Hypericum aethiopicum*). Doke and Vilakazi say of this plant that "roots [are] used to anoint the face" (1958: 508). The word also simply means 'face ointment'.

Also used for personal hygiene is the Aptenia (**Aptenia cordifolia*) with the name **umjuluka** (sweat) referring to the use of this plant as a deodorant. The same name (**umjuluka**) is used for the Sword-Leaf (*Casearia gladiiformis*).

On page 228 I gave the Zulu name **umphungwane** as possibly derived from *phunga* (drive off flies). It could also be a diminutive form of *umphungo* (disinfectant). Boon (2010: 150) gives the name **umphungwane** for *Craibia zimmermannii* (Sand Peawood), but makes no mention of its use in this manner and Coates Palgrave (1977: 312) is similarly silent.

On the other hand, Pooley (1993: 414) in giving the name **umgeza** (< *geza* (to wash)) for *Azima tetracantha* (Needle Bush or Bee-Sting Bush) states specifically that it is used as a disinfectant. Both Coates Palgrave (1977: 761) and Hutchings (1996: 237) note the use of this plant as a disinfectant and also say that sap from the plant is inserted into the wound after tooth removal.

According to Pooley (1993: 168) the leaves of the River Bean (*Sesbania sesban*) are used to produce soapy foam for washing the hands. The Zulu name given for this tree – **umkhumukhweqwe** – is an interesting one as it is derived from two ideophones: *khúmu* (of sudden action) and *khwéqwe* (of coming loose), a perfect description of dirt coming off the hands.

---

17. Cf. Coates Palgrave: "The bark provides a very strong fibre" (1977: 649).

More soapy lather is seen in the name **iphuphuma** (<*phuphuma* (overflow, bubble over, froth over)), a name given by Boon (2010: 288) for the Cape Holly (*Ilex mitis*). He says that the leaves are used to produce a soap-like lather. Coates Palgrave adds to this:

> If the fresh leaves are rubbed together in water a lather is produced; the Zulu bathe influenza sufferers in this and it is said that early Knysna woodcutters washed with it in the forest streams (1977: 492).

Pooley (1993: 378) gives the name **umgezisa** (what helps you wash) for the Rock Cabbage Tree (*Cussonia natalensis*), but neither she, Boon, Coates Palgrave nor Doke and Vilakazi mention any washing activities.

The case of the name **umgwenya** (*Harpephyllum caffrum*) is perhaps best discussed here. I have often wondered if this tree name was in any way linked to the noun *ingwenya* (crocodile), but for a long time could see no connection. I began to think about a possible link again when I read in Van Wyk and Gericke that "the bark is used to purify the blood in a person with 'bad blood', which manifest as pimples on the face" (2000: 232). To pursue the *umgwenya-ingwenya* link, we need to return to the notion of polysemy, last discussed at the end of Chapter 4 on semantics when I discussed the name **umkhondo**.

Doke and Vilakazi's dictionary (1958: 287) gives four different meanings for the noun *ingwenya*. The primary meaning is 'crocodile'. Meaning three, 'criminal, lawless person' is clearly linked to the first meaning in that both uses refer to something that attacks violently from hiding and is potentially dangerous to humans. Meaning four, 'submarine' is also clearly linked to crocodile in that both are cigar-shaped, hidden underwater, and potentially dangerous. It is meaning two that interests us specifically: 'ear of corn not properly threshed'. Such an ear of corn is not a hidden danger, but it is 'cigar-shaped' and the maize kernels that have not been threshed stick up from the otherwise plain silhouette of a corn husk in the same way that the plates and knobs stick up from a crocodile's skin, particularly on the tail. The illustrations on the next page should help illustrate this point. Now although a pimply face is not cigar-shaped, the pimples are indeed similar visually to the maize kernels sticking up from an ear of corn not properly threshed, and I venture to suggest that this is realised in the name **umgwenya**, given to a plant used to deal with pimply faces.

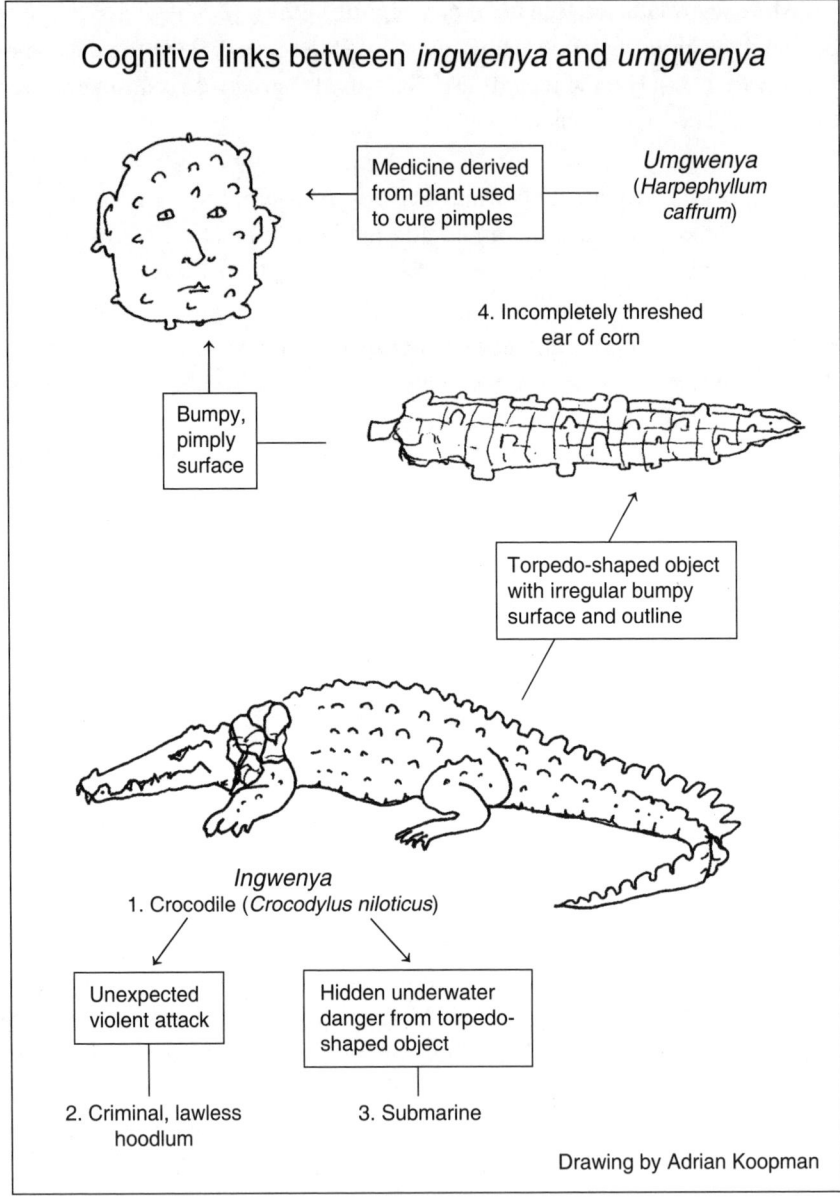

Cognitive links between *ingwenya* and *umgwenya*

Medicine derived from plant used to cure pimples

Umgwenya
(*Harpephyllum caffrum*)

4. Incompletely threshed ear of corn

Bumpy, pimply surface

Torpedo-shaped object with irregular bumpy surface and outline

Ingwenya
1. Crocodile (*Crocodylus niloticus*)

Unexpected violent attack

Hidden underwater danger from torpedo-shaped object

2. Criminal, lawless hoodlum

3. Submarine

Drawing by Adrian Koopman

Poisons, insecticides and glues

Two Zulu plant names refer to the use of the plant as an insecticide: **umfusamvu** and **umkhiphampethu**. Pooley (1993: 112) gives **umfusamvu** (what fumigates the sheep) for Cheesewood (*Pittosporum viridiflorum*). Despite the underlying meaning of this name, with its specific reference to sheep, Coates Palgrave (1977: 204) describes this plant as used to treat black gall sickness and red water in cattle. Hutchings (1996: 115) says roasted bark is used by black and white farmers for black gall sickness. One assumes she means black gall sickness in their cattle, rather than in themselves. She points out that saponins from the plant have antimicrobial properties.

For the Natal Laburnum (*Calpurnia aurea*), Pooley gives the name **umkhiphampethu** (what takes out the maggot) and says "infusion used against maggot-infested sores on cattle . . . crushed roots used against lice" (1993: 158). Doke and Vilakazi (1958: 394) give the underlying meaning of the name as 'what takes away vermin' and say the leaves and powdered roots of this plant are used as an insecticide. Pooley and other sources also give the name **insiphane-enkulu** (big insiphane) for this tree, where **insiphane** is almost certainly derived from *insipho* (soap).[18]

An example of a Zulu name referring to the use of a plant for glue is **inomfi** (bird lime), a name given by Boon (2010: 506) for the Wild Frangipani (*Voacanga thouarsii*), which, he says, is used for bird lime. Doke and Vilakazi (1958: 584) state that this word is used of a number of trees that provide bird lime, but Boon gives only this single species. Hutchings gives this name (in the form 'ino<u>n</u>fe') for *Erianthemum dregei*, saying "Zulu herders use the plant for catching birds" (1996: 80).

We have already noted in Chapter 6 the double meaning in the name **umhlalanyoni** (where the birds sit), the Natal Mistletoe (**Tapinanthus natalitius*), which could be a reference to the use of this plant as bird lime (which it is), or to the fact that it grows high up in a tree where the birds sit.

Fishing

I have recorded no Zulu plant names that refer to fishing. This is probably because fishing is an activity seldom practised by Zulu people, except in the far north of Zululand (Maputaland and Tongaland) on the boundary of the

---

18. The Zulu word *insipho* is adopted from the Afrikaans *seep*.

Zulu-speaking and Thonga-speaking areas. Although I have no data available to test this hypothesis, I would expect that the Thonga language of Mozambique has several plant names referring to fishing activities. Sources in *The James Stuart Archive* (Webb and Wright, 1976–2014) make occasional reference to fishing activities in Natal Bay (Port Natal, Durban) before the arrival of the first whites in the 1820s; and Bryant says that the noun *i-nDobela*, derived from the verb *doba* (to fish), refers to the "Tidal waters of the inner-bay or lagoon at Durban, so-called from the old-timed [*sic*] Native [*sic*] custom of fishing there" (1905: 113). Such fishing activities, however, have led to no entries in the Zulu botanomasticon, leaving one to conclude that fishing has never been a truly important social and cultural activity.

Miscellaneous

There are just a few left-over names to put in this inevitable category. Pooley (1998: 510) gives the Zulu name **umahlanganisa** (joining together) for *Urginea delagoensis*, a reference to the use of this plant as splints for bone fractures. Hutchings (1996: 39) notes that this plant is also used to protect animal skins against dogs, but this usage is not reflected in the name. It is not recorded in Doke and Vilakazi (1958).

Boon (2010: 496) gives the name **umvusankunzi** (what arouses the bull) for the Num-Num (*Carissa bispinosa*), but offers no explanation for the name. It is Hutchings who reports that the plant is:

> Reported to be used in traditional ceremonies by Swazi paramount chiefs to make sacrificial bulls fierce . . . The bulls are struck by a switch so that they will not be easy victims for the warriors, who have to overcome the animals with their bare hands (1996: 244).

I have not been able to ascertain whether the same is true of the bull-killing ritual that is part of the Zulu *uMkhosi woKweshwama* (the First Fruits ceremony) which, as reported in the media over the last few years, has been the focus of much protest from animal rights groups in South Africa and abroad. Hutchings also gives the name **umshayankunzi** (what hits the bull) for *Carissa bispinosa* and Boon adds the name **isibethankunzi**, which, although not recorded in Doke and Vilakazi (1958), is clearly from the verb *betha* (strike, hit) + *inkunzi* (bull). The name 'ibethamtunzi' that he also gives for this species would appear to be an inaccurate transcription from a hand-written source. This is an issue I shall discuss in the concluding chapter.

Doke and Vilakazi (1958: 312) record the name **umhlakaliso** as referring to a species of *Asclepiadaceae* with sweet-smelling flowers, the roots of which are given to dogs to make them hunt better. The name is derived from *hlakalisa* (make clever). Hutchings (1996: 251–253) gives extensive details about the various usages of eleven species of *Asclepias*, but the closest she comes to the usage recorded by Doke and Vilakazi is to note that the latex of an unidentified species of *Asclepias*, with the Zulu name **incohiba**, is smeared over eggs to prevent dogs stealing them. Doke and Vilakazi take this a little further by reporting **incohiba** as referring to a "small veld shrub, *Gomphocarpus*, used for poisoning dogs" (1958: 533). They give the same information for the variant **inqohiba**.

I end this sub-section and the chapter with an interesting link between Zulu plant names and social customs. Doke and Vilakazi (1958: 771) record the noun *insutha* (< *sutha* (be sated)) as having two different but related meanings. The first is simply the noun derived from the verb with the meaning 'satiety, eating to the full'. It is the second meaning that is particularly interesting. They state that as a plant name it refers to two species of grass, namely *Harpochloa capensis* and *Elionurus argenteus*, apparently worn in the hair to secure hospitality when travelling; in other words a traveller will achieve *insutha* (satiety) by wearing **insutha** in his or her hair! I can find nowhere in the literature other references to this curious, but delightful custom.[19]

The well-known traditional Zulu farewell is *Hamba kahle*! (Go well!). If you travel with a stalk of **insutha** in your hair, you are sure to go well.

---

19. Even Bryant's 1905 *Dictionary*, the 'feeder-dictionary' for Doke and Vilakazi, makes no mention of this.

10 The power in names

Fillet of a fenny snake,
In the caldron boil and bake;
Eye of newt, and toe of frog,
Wool of bat, and tongue of dog,
Adder's fork, and blind-worm's sting,
Lizard's leg, and owlet's wing, –
For a charm of powerful trouble,
Like a hell-broth boil and bubble.

In these words taken from Act IV, Scene 1 of Shakespeare's *Macbeth*, the three witches name out aloud the various ingredients they are cooking in a cauldron in order to create a powerful spell. When Hecate enters the scene and commends the three witches, she requires a song to be sung to finalise the magic:

And now about the caldron sing,
Like elves and fairies in a ring,
Enchanting all that you put in.

It is clear from this brief extract from *Macbeth* that the chanting, singing and naming out aloud of the ingredients are all essential parts of the magic. Materia such as newts' eyes, frogs' toes and dogs' tongues may well be important ingredients of the charm, but it is the utterances that go with the mixing of the ingredients that 'fixes' the magic. The word 'enchanting' is significant in itself, derived from the Latin *incantare* (to chant a spell).

Stuart Clark, writing on witchcraft and magic in early modern culture, explains the importance of 'utterance' in magic:

Above all, magic rested in the perceived power of words. Those who believed in this power think that there was a causal connection between words and their referents; they may simply have been exploiting the expressive capacities of language in a technological context heightened by ritual. Nonetheless, the assumption that words, simply by virtue of being uttered, had a mechanical power at least to assist in the causation or prevention of events seems to have been an intrinsic element of many of the procedures we noted earlier (Ankarloo, Clark and Monter 2002: 108).

This chapter will look at this power of words, particularly the power that resides in the Zulu names of plants, and at two aspects of this power: one, the power inherent in oral utterance, in other words the actual saying out aloud the names of plants and other ingredients of 'magic' in ritual chants and incantations (as with the three witches in Macbeth); the other, the power in names to aid the memory, in other words their value as mnemonics in a traditional oral society.[1]

The use of names in chants, incantations and spells

Ankarloo, Clark and Monter (2002: 100) give several examples of the importance of words and names in magic ritual in medieval Europe, emphasising the importance of both materia (things) and simultaneous utterance. One of their more unusual examples is that of a case in Zwickau in the early sixteenth century when a healer applied both fried onions and incantations to a man's head in order to cure him.

Stratton gives numerous examples of magic and orality from early Judaism and ancient Greece. She points out that sources as far ranging as the Hebrew Bible and Pliny the Elder see "conceptual magic as a form of 'performance utterance' – that is, words with the power to accomplish deeds" (2007: 9). As examples of "performance utterance" Stratton relates two stories concerning the Greek heroes Jason and Odysseus (2007: 27). According to her, Jason seduced Medea with a ritual he had learned from Aphrodite: he tied a bird to a wheel and uttered prayers (*litas*) and incantations (*epaoides*). The same Greek word – *epaoidē* – appears in Stratton's story of Odysseus. One day, when out hunting, Odysseus was gored by a wild boar. His uncles stanched the wound

---

1. This chapter is based on my article 'The interface between magic, plants and language' *Southern African Humanities* 25 (2014).

with an *epaoidē*. The word is usually translated as 'charm or 'incantation' and is etymologically related to singing. In other words, Odysseus' uncles <u>sang</u> a cure into the wound.

When I told the story of Odysseus to Ndela Ntshangase he immediately recalled incidents from his youth when he and his friends, throwing stones at each other accidentally hit one of the young chickens in the yard and it would fall over as if dead. The frightened boys would gather round it and sing *'Vuka Mangozi, kade sidlala! Vuka Mangozi, kade sidlala!'* (Wake up, Mangozi,[2] we were just playing) until (hopefully) the young bird would recover and get up.

In other parts of the world, the oral part of magic is equally important. In a recent article on the vanishing languages of the world, Russ Rymer talks about the dying language Aka, spoken in Arunachal Pradesh, a rugged north-eastern state of India. He tells of 25-year-old teacher Pario Nimasow, who brings to the author a bag of ritual items: a tiger's jaw, a python's jaw "and other objects of a shaman's sachet":

'My father was a priest,' Nimasow said, 'and his father was a priest.' And now?, I asked. Was he next in line? Nimasow stared at the talismans and shook his head. He had the kit, but he didn't know the chants; his father had died before passing them on. Without the words, there was no way to bring the artifacts' power to life (2012: 77).

So far, we have looked only at the power of words in actual utterance, the voicing out aloud ritual incantations that may also contain the names of the ingredients of a charm. But part of the power of a name may also reside in its underlying meaning. Victor Turner (1977: 23) explains that among the Ndembu people of north-western Zambia, collecting herbal medicines from a tree must always be accompanied by invocations to the trees. In addition, he explains, there is power inherent in the names of the trees themselves:

The *muhotuhotu* tree (*Canthium venosum*) is used for medicine 'because of its name'. Ndembu derive this from *ku-hotomoka*, 'to fall suddenly, like a branch or fruit' (1977: 26).

---

2. Mangozi is a 'nonce-name' based on the noun *ingozi*, here with the meaning of 'accident'. The name could be translated roughly as 'the one hit by accident'.

A similar point is made by Sir James Frazer of *Golden Bough* fame in talking about the Toradja people of Indonesia:

> In the fire [the Toradja rain-doctor] burns various kinds of woods, which are supposed to possess the property of driving off rain; and he puffs in the direction from which the rain threatens to come, holding in his hand a packet of leaves and bark which derive a similar cloud-compelling virtue, not from their chemical composition, but *from their names*, which happen to signify something dry or volatile [my emphasis] (2002: 69).

Ndela Ntshangase gives two similar examples involving plants and their Zulu names: the plants *Acridocarpus natalitia* and *Spermacoce natalensis* both have the name **umabophe**, a name based on the verb *bopha* (to arrest, to stop, to bring to a halt). The plants are used in 'binding' spells, commonly used to 'freeze' an intruder in his steps. The relationship between the underlying meaning of the name and its function is immediately obvious. In addition to this 'powerful meaning', says Ntshangase, when medicine containing this plant is sprinkled around the perimeter of the homestead, the person doing the sprinkling chants '*Mabophe wami, mabophe wami, bopha noma ubani ozokwenza ububi la*' (Mabophe of mine, Mabophe of mine, halt whoever it is who is intending to do evil here), repeating this until all the medicine has been sprinkled. The plant *Osyridicarpus natalensis* is used in the same way. Its Zulu name is **umalala**, a name based on the verb *lala* (lie down, go to sleep). Ntshangase says that medicine containing the plant is used as an *intelezi* (sprinkled medicine) when a wedding or other occasion is imminent and the host does not want the guests to fight among each other. Sprinkling the umalala medicine around the venue, he chants '*Malala wami, Malala wami, ngicela laba banti bangalwi*' (Malala of mine, Malala of mine, I request that these people do not fight), also repeating this refrain until all the medicine has been sprinkled.[3]

Ntshangase's example here makes an interesting contrast with 'binding' spells used in Roman times. Stratton describes such a spell as

> constraining a person by 'binding' him or 'nailing him down' with the use of an image or other proxy, often a folded piece of lead, on which the victim's name had been engraved, pierced with a nail.

---

3. Personal communication.

. . . This 'curse tablet' was then placed in a significant place, like a well or a grave [a liminal place: one on a boundary] . . . These spells were enlisted most often in situations of competition – competing lovers, athletes or business rivals (2007: 41).

Note that in Stratton's example, the name is written on a piece of lead whereas in Ntshangase's examples, the names are uttered aloud. Note also that in Ntshangase's examples it is the name of the plant containing the binding power, whereas in Stratton's example the name of the intended victim is used. Naming the intended victim of an evil charm is a worldwide practice and we have already noted in Chapter 8 how an *umthakathi* uses the plant **uqhume** (**Hippobromus alata*).

The notion of power inherent in certain plant names applies also to plants used in medicinal preparations. The following examples have already been discussed in Chapter 7: **umsindandlovana** (*Ormocarpum trichocarpum*), **ikhaphanyongo** (*Melanthera scandens*), **ubangalala** (*Rhynchosia* species), and **umakuphole** and **icishamlilo** (both used for *Pentanisia prunelloides*). In each case the effectiveness of the medicine itself is greatly strengthened by the name of the plant and it is of course essential for the doctor using the plant to tell the patient its name. Thus a person with a particularly bad case of a certain illness will be greatly cheered when the doctor says he is treating the problem with **umsindandlovana** (even an elephant will recover) and a patient presenting with liver problems will be encouraged when he is told he has been given medicine containing **ikhaphanyongo** (what drives away the gall). Similarly, a man with erectile problems will have faith in a medicine containing **ubangalala** (what will help with intercourse) and a person with a high fever will be heartened knowing she is being cured with medicine based on **umakuphole** (let it cool down) and **ichishamlilo** (put out the fire).

The power of plant names as mnemonics

Walter J. Ong states that "an oral culture has no texts" and then asks the question: "How does it get together organized material for recall?" He answers his own question as follows:

The only answer is: Think memorable thoughts. In a primary oral culture, to solve effectively the problem of retaining and retrieving carefully articulated thought, you have to do your thinking in mnemonic patterns, shaped for ready oral occurrence. Your thought must come into being in heavily rhythmic, balanced patterns. In repetitions or antitheses, in alliterations and assonances, in epithetic and other formulary expressions, in standard thematic settings . . . in proverbs which are constantly heard by everyone so that they come to mind readily, and which themselves are patterned for retention and ready recall (1982: 34).

Ong's thoughts here apply particularly to oral poetry, but are certainly valid when applied to botanical nomenclature in an indigenous knowledge system. Indigenous knowledge systems do indeed occur in primary oral cultures and users of such systems have to remember details of identification and use without the benefit of notebooks, reference books, dictionaries, or any kind of archived written material. It makes sense, then, that names used, for example, in a system of indigenous (oral) botanical nomenclature have strongly mnemonic characteristics.

Any plant name where the underlying (lexical) meaning is different from the denotative meaning has mnemonic qualities. To see this clearly, it is perhaps necessary to compare such names with names where there is no difference between the two types of meanings, or no discernible underlying meaning at all.

Let us look at the Zulu plant names **umunga**, **umvumvu**, **intovane**, **isifithi**, **umncuma** and **umtholo** to take just six at random.

The plant name **umunga** has two referents – *Acacia karroo* and *Acacia natalitia* (Boon 2010: 178, 182). Leaving out *A. natalitia* for the moment, we note of *A. karroo* that like any acacia it is possessed of considerable spines and, in addition, is a good source of twine. Neither of these salient features can be discerned in the underlying meaning of the name. Indeed, it could be said that the name *umunga* has only referential meaning. There is nothing about the name that helps anyone to remember what it looks like, or what it is used for. Compare this name (**umunga**) to the name **umngampondo** for the tree *Acacia grandicornuta* (Horned Thorn). Here the underlying meaning is 'acacia with horns',[4] a link between lexical meaning and appearance of the tree that is decidedly mnemonic. Then again, the name **umunga** says nothing

---

4. An underlying meaning in the Zulu language that has also certainly been at the base of the specific name *grandicornuta* and the English common name Horned Thorn.

about the use of the bark from this tree to produce twine. There is no mnemonic link. Compare this to the Zulu name **usingalwesalukazi**, given by Pooley for *Gomphocarpus physocarpus*, who says "the stems are used for fibre" and in a rare acknowledgement of an apposite underlying meaning states the "Zulu name refers to the fibre used for sewing i[z]idwaba, [the] leather skirts of . . . women" (1998: 168, 546). Here the Zulu name (lit. thread of the old lady) is a reminder of an essential use of the plant.

The plant name **umvumvu** is one of several names for the Pigeonwood (*Trema orientalis*), but it is best known as a name for the White Stinkwood (*Celtis africana*) of which Boon says "wood reputed to have magical properties, used with crocodile fat against lightning" (2010: 66). Again, there is no indication of these 'magical' properties in the name **umvumvu**, which cannot be considered to have a separate underlying meaning. In other words, there is nothing about the meaning of **umvumvu** that would help one remember its function as a protection against lightning. Compare this to the several names discussed in Chapter 8 under the heading 'Protection against the weather' where we discussed names like **isiwisa** (cause [hail] to fall), **umkhombazulu** (point to the sky) and **ubani** (lightning). What better way can there to be to remember that a particular plant is efficacious against lightning than to call that plant 'lightning'?

Boon says of the tree *Smodingium argutum*, aptly named Pain-Bush in English, that the "sap raises blisters & causes hospitalization in severe cases". He goes on to talk of its "terrible reputation" (2010: 286). One might expect a plant with such a reputation and such side effects to have a name that would reside in the memory. And yet the Zulu name **intovane** has no separate underlying meaning and the name can be considered, mnemonically speaking, a failure. Compare this to the word *umthelelo*, which Doke and Vilakazi (1958: 790) say refers to a poisonous concoction placed on the ground to kill by witchcraft. When this is also used for the tree *Ochnea arborea* (Boon 2010: 360), the link between the primary meaning of the noun, its use as a plant name and the way the plant is used in society is obvious.

Isifithi is recorded by Boon as a name for the tree *Baphia racemosa* of which he notes "not for hut building because [it is] reputed to bring bad luck" (2010: 146). As with **intovane** above with its "terrible reputation", we might expect this quality of bad luck to be memorable in the name. But **isifithi** has no underlying meaning at all and therefore no such mnemonic qualities. Then again, I am unable to find any tree name in the Zulu botanomasticon[5] that has an underlying meaning referring to avoidance of use of its wood because of potential bad luck.

According to Boon the bark of the tree *Olea woodiana* is "used as an appetite stimulant and a nerve tonic" (2010: 478). You would never know this from the Zulu name **umncuma**, which has no separate underlying meaning. Compare this to the plant names **umnandi** (lit., tasty) and **umfan' ozacile** (lit., the boy is thin) for *Syncolostemon parviflorus* and *Kohautia amatymbica* (Pooley 1998: 188 and 202). Pooley says *K. amatymbica* is used "to improve appetite in infants" and *S. parviflorus* is "used in traditional medicine to treat loss of appetite". It may be labouring the point if I say **umnandi** suggests mnemonically, through its connotative meaning 'this should tempt the child to eat because it is so tasty', a link between the plant and one of its known functions, with **umfan'ozacile** making a similar link through its connotative meaning 'a thin boy needs fattening up through feeding him tasty foods'.

Earlier in this chapter I mentioned the plant names **ikhaphanyongo** (what takes out the gall), **ubangalala** (what will help with intercourse; lit., what causes [one] to lie with [another]), **umakuphole** (let it cool down) and **icishamlilo** (put out the fire), suggesting that the underlying meanings of these names help the patient believe that he or she is getting the appropriate medicine for the illness. It is equally likely that the underlying meanings of these names help the doctor to remember what plant to use in particular circumstances. Like the written labels on medicines that come from a modern pharmacist that not only identify the medicine, but remind the patient to 'take three times a day for stomach cramps', names like **ubuhlungu-bemamba** (pain of the mamba), **umahayiza** (rave, be hysterical) and **uvimbukhalo** (what blocks up the loins) help the *inyanga* to remember that these are used for treating snakebite, hysteria and diarrhoea respectively.

Consider also how underlying meanings can help a traditional healer to remember where to find the plants he needs. Think of an *inyanga* who is starting to run out of *Ranunculus multifidus*, used for coughs, urinary tract complaints and venereal ulcers. He does not have think very deeply about where to find more stock – the name **uxhaphozi** (< *ixhaphozi* (swamp)) tells

---

5. The term 'onomasticon' is used by onomasticians to refer to a specific lexicon of names for a specific purpose such as, for example, the 'English anthroponomasticon', the names commonly used for personal names among English-speakers. I have coined the term 'botonomasticon' to refer to a list of plant names common to a particular language.

him immediately. The same would apply to anyone needing to replenish stocks of *Burchellia bubalina* with the Zulu name **umvuthwemfuleni** (what ripens in the river) and *Ochna natalitia*, known in Zulu as **umilamatsheni** (what grows among rocks). Like *Ranunculus multifidus*, *Dissotis canescens* (Pink Marsh Dissotis) grows in marshy areas (Pooley 1998: 410). Remembering this is slightly more complicated for an *inyanga* as he first has to make the mental links between marsh and frog, and marsh and crab: two Zulu names for this plant are **imfeyesele** (sweet-reed of the frog) and **imfeyenkala** (sweet-reed of the crab).

Once the underlying meaning of a habitat has stimulated a traditional collector of plants into finding the right local environment, he or she then needs to know what is actually being looked for. Here an underlying meaning that describes the appearance of a particular plant is an effective mnemonic. In Chapter 6 we looked at a number of plants with underlying meanings describing appearance such as **umpondonde** (long horns), the Tree Aloe (*Aloe barberae*); **uvemvane olubomvu** (red butterfly) the Little Russet Pea (*Argyrolobium tuberosum*); and **isinambuzane** (insect) the Lesser Moth-Fruit Creeper (*Sphedamnocarpus pruriens*). Occasionally, as we noted, the mnemonic is in the form of a metaphor that links plant to a body part of a particular animal as in **indlebeyemvu** (sheep's ears), the Sheep's Ears Everlasting (*Helichrysum appendiculatum*); and **unyawolwenkuku** (hen's foot), the Stalk-Flowered Pelargonium (*Pelargonium luridum*).

In effect, scientific botanical names descriptive of the appearance or structure of a particular plant have the same mnemonic qualities as the Zulu names in the paragraph above. Hundreds of examples could be gleaned from any botanical reference book and I take the following three plants randomly from Pooley (1998): *Ceropegia crassifolia, Brachystelma circinatum* and *Craterostigma nanum*. Pooley explains the first: *Ceropegia* is derived from *keros* (wax) and *pege* (fountain) as "Linnaeus thought the flowers looked like a fountain of wax" (1998: 172), while *crassifolia* is derived from *crassus* (thick) and *folia* (leaves). The genus name in the second example is derived from *brachy* (short) and *stelma* (crown), while the specific epithet is derived from *circinatus* (coiled inwards from tip) (1998: 172). In the third example, *Craterostigma* is based on *krateros* (mouth of volcano), a reference to the cup-shaped stigma, while *nanum* means 'dwarf' (1998: 194). The English vernacular Mole's Spectacles is a lovely example of whimsy and metaphor.

Sometimes the mnemonic works the other way around: a characteristic of the plant helps remind a traditional healer what its name is. This would certainly be the case for almost all plants that exude sticky white latex reminiscent of *amasi* (sour milk). This visual reminder quickly brings to mind names like **inkamasane** (what squeezes out a little maas), **inkalamasane** (what weeps maas) and **amasethole** (maas of the heifer), the Zulu names for *Euphorbia bupleurifolia*, *Euphorbia natalensis* and *Sideroxylon inerme* respectively.

In the quotation from Ong at the beginning of this section on mnemonics, he asked how one remembered things in a primary oral culture and his answer was "think memorable thoughts". In the primarily oral traditional Zulu society, one might ask how traditional healers remember what plant is what, where to find them and how to use them. The answer might well be 'give them memorable names'.

11 Conclusion

It has also been emphasised that the various meanings of African names throw light upon the whole traditional culture, and thus they may serve in reaching a deeper understanding of the people, their ideas and their way of life (Saarelma 2009: 195).

Saarelma, in this quotation, is talking specifically about anthroponyms, the names of people, but the point she makes is equally valid for what I have dared to call 'botonyms' – the names of plants. In this concluding chapter we shall try to sum up the links between plants, the names of plants, the people who name and use plants, the language that the people use to name plants, and the culture that embraces all of these.

First we need to go back to the idea of 'systematic naming', or 'naming systems'.

All naming takes place within a particular naming system

Whether we are talking about the naming of people, the naming of places, or the naming of plants, such naming takes place within a complex system. As in any system, there are the constituents (the several parts) of the system, the different and hierarchical categories into which these constituents are sub-divided, and the dynamic links between the constituents and sub-constituents at all levels of the hierarchies. In <u>any</u> naming system the four major constituents are (1) the names, (2) the named, (3) the namers and (4) the context in which they are named. These four constituents apply equally to the scientific system of botanical nomenclature and the system of plant naming in the Zulu culture described variously in this book as part of an 'indigenous knowledge system', as 'folk taxonomy' and as 'traditional ecological knowledge'.

Let us look again briefly at each of these.

The names

As we have seen, all names are nouns, linguistic items subject to the grammatical and phonological rules that govern each language. In botanical nomenclature, the language is always Latin and a set of strict rules govern the way in which these names are formed. In the Zulu botonymic system, the language is, of course, always Zulu, although we noted at the end of Chapter 1 a few plant names derived from English or Afrikaans. All words, of all kinds, in all languages have a structure or shape and the structure of Zulu plant names was thoroughly analysed and described in Chapter 3. Apart from 'nonsense words', all words in all languages have meaning and we noted in Chapter 4 and elsewhere that there are different types of meaning, and often different layers of meaning, in one word. We have noted that the primary meaning of words is a referential meaning, that layer or type of meaning which links language to 'real life' entities. In the case of plant names of all types this referential meaning identifies a species or a group of plants.

The underlying meanings link the plant to various other things: the physical nature of the plant, its habitat, the part of the world where it is found, tributes to humans involved in one way or another with botany, and in the case of the Zulu plant names to the way the plant is used in human culture: medicinal, as charms and for various practical purposes. These links are examples of the 'dynamic links between the constituents and sub-constituents at all levels of the hierarchies' mentioned earlier.

The named

The well-known Scottish doyen of onomastics, Bill Nicolaisen, once said humans should be known as *Homo nominans* – 'man the namer' (Nicolaisen 1974) – from the tendency of humans to give names to almost every facet of their world: to each other, to the places they live in, to geographical features, to manufactured products, to literary characters and to a wide variety of living things: wild and domestic animals, birds, insects and, indeed, plants. These are, as we noted in the introductory chapter, not necessarily all what linguists would call 'proper names', but they are names nonetheless. In this book, the named entities have been (apart from a few names of people and places for comparative purposes) plants of all kinds: trees, bushes, shrubs, herbs, vines, fungi and so on. We have also noted the names of medicines and concoctions, themselves derived from plants or combinations of various parts of plants.

In the system of botanical nomenclature the named entities are specifically taxonomic categories: classes, families, genuses and species. This is not quite the case with vernacular names where the link between species and name, as we have seen, is not so hard and fast. But we have noted that both in botanical nomenclature and with vernacular names, a name may refer to the species as a whole (or to a group of plants perceived to be related in some way) or to an individual specimen. There is a distinction in reference in the two statements 'Acacia sieberiana grows extensively in the thornveld between Pietermaritzburg and Durban' and 'That is a fine Acacia sieberiana you have growing there in your garden', just as there is a difference between 'The carnation is often found in men's buttonholes, especially at weddings' and 'Please remove the carnation from your buttonhole immediately!'

The namers

In Chapter 2 we noted that in the botanical system of nomenclature the act of naming can often be precisely defined, with the namer known, the date of naming known and the reason for naming explained. In a system like the Zulu system of plant naming, and indeed in most vernacular naming of plants, these details are not known. We can still talk of the 'namers' though, but to refer to the people who use plant names rather than those who actually coin them. In Chapter 5 we looked at five different classes of people who use plants and so need to describe and 'name' them: the traditional healers, the shamans, the witches, the 'heaven-herds' and the ordinary 'man-in-the-street'. In both Western and African societies there are those members of society with a great deal of knowledge about the world of plants, and consequently have a large lexicon of plant names stored in their brains, and those who seldom deal with plants and consequently have a much more limited lexicon.

To illustrate, common vernacular names are used by all sectors of the population in a particular language area. Mother-tongue English speakers are perfectly familiar with plant names like 'oak', 'fir', 'pine', 'rose', 'daisy', 'marigold', 'parsley', 'thistle' and 'dandelion'. We would expect to find the same with common plant names in Afrikaans and Zulu, French and Sotho, Tswana and German.

But I would imagine that one would have to be a keen gardener, or an amateur or professional botanist, to be able to tell the difference between a Natal Thesium, a Green Wood Orchid and a White Hairbell.[1] All English mother-tongue speakers know the meanings of the words 'five', 'finger', 'grape', 'forest' and 'glossy'. But how many can tell the difference between the Five-Finger Grape (*Rhoicissus digitata*), the Forest Grape (*Rhoicissus revoilii*) and the Glossy Forest Grape (*Rhoicissus rhomboidea*)? How many non-specialists would even know that the phrases 'honey locust' and 'ribbed currant' refer to trees?[2] Or that 'white trumpets', 'string of stars' and 'two-day cure' refer to various herbs?[3]

The context

All naming takes place within a context of some kind: cultural, political, social, historical, economic and others. In the system of botanical nomenclature, the context can loosely be described as 'professional'. Botanical names tend to be used by scientists, notably and obviously botanists, and other professionals such as horticulturalists, nurserymen and educators, especially at a tertiary level. When we look at Zulu plant names, the context can loosely be described as 'cultural'. The use of plants in Zulu society and many of the underlying meanings of their names are linked strongly to Zulu traditional cultural beliefs. We saw this particularly in Chapters 7, 8 and 9, which dealt with plants used medicinally, plants used as charms and plants used for a variety of practical purposes. To understand the relationship between a particular underlying meaning of the name of a certain plant and the plant itself, we have found it necessary to understand, for example, traditional beliefs about the nature of lightning, the role of the 'heaven-herd' and the plants that may be used to counter various threats of adverse weather. With other plants it has been necessary to examine the role played by the *umthakathi* in Zulu society.

In Chapter 8 I quoted Hammond-Tooke as saying that the "integration and smooth functioning . . . [of society] . . . are ensured by the mechanisms of political institutions and the supernatural sanctions of religion" (1962: 263). I noted that when a society functions smoothly, young men court young maidens.

---

1. *Thesium natalense, Bonatea speciosa* and *Dierama luteoalbidum* respectively.
2. *Gleditsia triacanthos* and *Searsia pallens* (Boon 2010: 138, 280).
3. *Chascanum hederaceum, Heliotropium steudneri* and *Senecio serratuloides* (Pooley 1998: 180, 178, 328).

If successful, they marry them; if things go well, children are born; and so the cycle continues. But there are potential problems in the smooth functioning of such a cycle. A man may not be able to attract a potential bride; a young bride may find it difficult or impossible to conceive. For these problems there are plant-based love charms and medicines that can be used; and as we saw in Chapter 8, the plants so used frequently have names that refer specifically to these threats.

Infertility and inability to conceive are just two of many other threats that face the harmonious functioning of society. There is the malice of *abathakathi*, the jealousy of neighbours and co-wives; there is bad luck and illness; crops may fail and livestock fail to reproduce; lightning may strike, hail may fall and snakes might bite. All of these, and more, are reflected in the underlying meanings of plant names.

It is without doubt that, as Saarelma says in the quotation at the beginning of this chapter, "the various meanings of African names throw light upon the whole culture" (2009: 195).

Challenges for the future

One of the major themes running through this book has been that Zulu names for plants are part of an indigenous knowledge system, operating within a primarily oral culture. In such a knowledge system, and in such a culture, knowledge is held in the 'group memory' and transmitted orally from generation to generation. In Zulu society most botanical knowledge was, and still is, held by the *izinyanga*, the equivalent of the 'apothecary-doctors' who held the same knowledge in the first and second millennia of the current era. In traditional Zulu society aspirant traditional healers followed in the footsteps of a father, grandfather or uncle, learning while serving as apprentices. Acquiring the knowledge of the elders took many, many years. In the 1990s a group of concerned scholars became aware that young men were no longer keen to follow in the footsteps of their elder kin who were *izinyanga* and that as a consequence the pool of traditionally held knowledge was starting to diminish. It was this group of scholars, mainly botanists and Zulu-language experts, who formed the first working group of what would become the Zulu Botanical Knowledge Project (ZBKP).

Recognising that in existing literature errors were numerous, both in the spelling of Zulu names and assigning of them to particular species of plants, that many plants did not have Zulu names recorded even though they grew extensively in Zulu-speaking areas and that many plants had several names, often because of regional variations, the initial aim of the ZBKP was to employ Zulu-speaking investigators. They would cover the whole of KwaZulu-Natal, interviewing both *izinyanga* and *izangoma* as well as professional plant collectors, collecting specimens of plants and recording the various names used as well as any other information of plant usage. It was an ambitious undertaking, and the working group decided to start with a pilot project with Mkhipheni Ngwenya of the Natal Herbarium at the Botanical Gardens in Durban, with a small group of helpers, investigating plant names in only three carefully selected areas of KwaZulu-Natal: the Ngoye Forest near Mthunzini north of the uThukela River, the eNtimbankulu Forest on the South Coast near Port Shepstone and the Lotheni area near Bulwer in the foothills of the Drakensberg. This pilot project resulted in a publication in 2003.[4] Unfortunately, after this publication the ZBKP lost steam, mainly as a result of funding difficulties. Since 2003 a number of attempts have been made to revive the project, but at the time of writing, nothing has been finalised.

A dictionary of Zulu plant names, based both on names recorded in the existing literature as well as extensive modern field research, especially one recording regional variations, would go a long way towards reviving the gradually fading traditional botanical knowledge of the Zulu people. Such a dictionary would make an admirable reference book, but its sheer massiveness would preclude it being used in 'lighter' publications, such as Zulu-language textbooks on botany for schools and tertiary institutions, or field guides for the interested layperson. And in the actual process of recording the traditional oral knowledge are all the attendant problems that come when one type of knowledge system is replaced by another.

I made some remarks in Chapter 2 of this book about pitfalls that loom when an oral system of knowledge is replaced by a written one. One is that in oral knowledge there is no such concept as a single standard form, whether a eulogy composed for a king, the praises of a clan or a plant name. I have given the several potential forms of **umfana kaNozihlanjana** (*Stylochiton natalensis*) and shown that morphological variation can produce variants like **isankuntshane** and **isinkuntshane**, **inozane** and **inonzane**, **isibhaha** and **isibhane**, and **isaqathe**, **isanqante**, **umanqante** and **inqante** – all perfectly

---

4. The details of the aims of the project and members of the working group can be found in the project's first (and so far only) publication (Ngwenya, Koopman and Williams 2003).

valid names in an oral culture. A Euro-Western style publication would ask 'Which of these forms is the correct one?' and perhaps arbitrarily select one for publication while consigning the others to oblivion. And then again an oral culture has no spelling. Once oral forms become written forms, spelling (orthography) becomes an issue. Rules of orthography change and there may be differences between orthographies in different countries as still affects writers of English who must decide between American or British spelling; or as for many years the spelling of words in Sotho differed markedly in the systems used in Lesotho itself and in South Africa. With spelling of written words come the vexing questions of capitalisation, treatment of noun class prefixes in Bantu languages, and the issue of writing phrasal names as compound words, as hyphenated words or as separate words. All these issues were discussed in Chapter 2.

It is illuminating to see how various publications on mammals, birds, amphibians, reptiles, fish, insects, etc. treat vernacular names from South African Bantu languages. By and large they ignore these languages completely. The following is a very brief survey. Note that all the books mentioned below give the appropriate scientific binomial name for each species mentioned.

Survey of vernacular nomenclature in the natural sciences

Smithers' *Mammals of the Southern African Subregion* adds English and Afrikaans vernacular names only with a brief paragraph for each species discussing the use of alternative English and Afrikaans names. Thus in describing *Ictonyx striatus*, and giving the vernacular names Striped Polecat and Stinkmuishond (1983: 441), it is noted the animal has also been known as 'Cape Polecat', 'skunk' and 'zorilla'. The only mention of Zulu comes in discussion of the vernacular names of the impala and the nyala. Of the nyala (*Tragelaphus angasii*), Smithers says "The colloquial name for the species originates from the Zulu name for the species, inxala [*sic*]" (1983: 671). The Zulu name for the nyala is *inyala*; *inxala*, according to Doke and Vilakazi (1958: 614) is the Rhebok (*Cervicapra lalandii*).[5] Smithers says of the colloquial name 'impala' that it "was most probably anglicized from the Tswana name *phala*, although it has been claimed that it is derived from the Zulu name for the species, iMpala" (1983: 647), a statement I find totally mystifying.

---

5. Although Walker (1981: 179) assigns the name *inxala* to the Reedbuck (*Redunca arundinum*), creating the same kind of confusion that we expect when Zulu names in one source are compared to the same name in another source.

Wagner's *Frogs of South Africa* (1986) adds only English colloquial names to the scientific binomials as is the case with Skaife's *African Insect Life* (1979). Clinning's *Southern African Bird Names Explained* (1989) has a title promising (at least it did to me when I first saw the book) that it would provide a wealth of information on bird names in various South African vernaculars. Alas, his explanations are restricted to the scientific names, although it must be said that an (unexplained) English vernacular name goes with each entry.

Patterson's *Reptiles of Southern Africa* (1987) restricts vernacular names to English and Afrikaans, giving variants where appropriate and occasionally mixing English, Dutch and Afrikaans together as in "Rainbow or Blue-Tailed Kopje-skink, Bloustert-koppieskink" with the arresting scientific name *Mabuya quinquetaeniata margaritifer* (1987: 50).

Clive Walker's *Signs of the Wild* is the earliest book of which I know that talks generally about wildlife and attempts to give names in all of South Africa's vernacular languages. So in the entry for the Antbear (*Orycteropus afer*), we find in addition to the expected Afrikaans name *aardvark*, a name in Shona (*Sambane*), Ndebele (*Isambane*), Zulu (*Sambane*), Shangaan (*Xombana*), Venda (*Thagalu*) and Transvaal Sotho (*iThakadu*) (1981: 15). A careful scanning of this list shows that the Ndebele name *isambane* has an initial vowel, whereas the almost identical Zulu language has the initial vowel missing (*sambane*). Walker does this consistently throughout the book – removing the initial vowel that is – from every Zulu animal name, despite the fact that all Zulu nouns have an initial vowel. I cannot say why he should have decided to do so. Apart from that almost all Zulu names for animals are accurate, although there are some very curious entries: 'nehi' for the Aardwolf[6] (*Proteles cristatus*); the spelling 'Nkhonhoni' for *inkonkoni* (the Blue Wildebeest *Connochetes taurinus*); giving only the female name (*imbabala*) of the bushbuck (*Tragelaphus scriptus*) and leaving out the name of the male (*unkonka*); spelling *indluzele* as 'Nduluzele' (Red Hartebeest *Acelaphus buselaphus caama*); and more. Nonetheless, Walker's *Signs of the Wild* is a milestone in the recognition of 'African' names.

My earliest edition of *Roberts Birds of South Africa* is the fourth, revised by McLachlan and Liversidge in 1978; i.e., three years before Walker's *Signs of the Wild*. For some of the species of birds in the book, one or two 'Native

---

6. The Zulu name is *isingci*.

names' are given.[7] When my colleague on the Pietermaritzburg campus of the then University of Natal, Gordon MacLean, indicated in the mid-1980s that he was working on the fifth edition of *Roberts' Birds*, I asked him if it was possible to do considerable work on the Zulu vernacular names as my own research showed that a great number of valid Zulu names were missing, and many of those recorded in the fourth edition were incorrectly assigned or poorly spelt. He agreed, and we later spent many hours incorporating my work on Zulu bird names into the fifth edition. MacLean then found experts on bird names in the other 'African' vernacular languages and enlisted their help as well, with the result that the 'MacLean edition' of *Roberts' Birds of Southern Africa* (1985) contained very credible listings of bird names in Zulu, Xhosa, Tswana, Southern Sotho and other southern African languages.

And then we come to publications of botanical reference works and as we have seen in the review of sources in Chapter 2 of this book, botanical writers began very early to include African vernacular names for plants. I have used lists of Zulu plant names from Bryant (1909, re-published 1970), Bews (1921), Gerstner (1938, 1939 and 1941), Smith (1966), Palmer and Pitman (1972), Moll (1981), Roberts (1990), Pooley (1993, 1998 and 2003), Hutchings (1996), Boon (2010), Van Wyk et al. (2011) and Paulsen et al. (2012).[8] These publications spanning over 100 years have used Zulu plant names for their referential or denotative meanings, in other words as an aid to identify certain plants. Only very occasionally does a note on the underlying connotative meaning creep in. To my knowledge, Ngwenya et al. (2003) is the only publication that focuses strongly on the underlying meaning of the Zulu plant names presented.

Conformity in the provision of vernacular names

Despite over 100 years of collecting Zulu plant names and including them in publications, there remains a problem in the provision of Zulu names for various species. I provide the following example from Boon, chosen completely at random (2010: 270–86). He enumerates 36 different species of *Searsia*, previously known as the genus *Rhus*. Each one, even the varieties of single species, has an English vernacular name and an Afrikaans vernacular name,

---

7. In the Introduction these are referred to as "Native tribal names" (1978: xvii). There is no indication where these came from.
8. Published in 2012, but based on Zulu plant names collected by Blessing, 1903–1904.

and sometimes more than one of each. See the full table where I have given the scientific name, and the English and Afrikaans vernaculars. Where possible I have given the Zulu vernaculars, with 'NZN' to indicate "No Zulu Name", leaving out Xhosa and Swazi names. Of the 36 species enumerated, 22 (61%) have the Zulu name **inhlokoshiyane**. One species (*S. chirindensis*) has four Zulu names: a qualified form of **inhlokoshiyane** (**inhlokoshiyane enkulu** or big inhlokoshiyane), **umhlabamvubu**, **ikhathabane** and **umdwendwelengcuba**. The name **umhlabamvubu** is also assigned to *S. dentata*. One species (*S. gueinzii*) also has the Zulu name **umphondo**. So although no English or Afrikaans vernacular is used for more than one species of Searsia, the Zulu name **inhlokoshiyane** is used for 22 species.

Searsia and vernacular names.

Scientific name	English name(s)	Afrikaans name(s)	Zulu name(s)
S. acocksii	Pondo Climbing Currant	*pondoranktaaibos*	NZN
S. carnosula	False Nana-Berry	*basternanabessie*	NZN
S. chirindensis	Red Currant	*bostaaibos*	**inhlokoshiyane-enkulu** **umhlabamvubu** **ikhathabane** **umdwendwelengcuba**
S. crenata	Dune Crow-Berry	*duinekraaibessie*	NZN
S. dentata	Nana-Berry	*nanabessie*	**inhlokoshiyane** **umhlalamvubu**
S. discolor	Grassveld Currant	*grasveldtaaibos*	NZN
S. divaricata	Rusty-Leaf Currant Mountain Kuni-Bush	*roesblaartaaibos* *bergkoenibos*	NZN
S. dracomontana	Drakensberg Dwarf Currant	*drakensberg-dwergtaaibos*	NZN
S. fastigata	Broom Currant	*besemtaaibos*	**inhlokoshiyane**
S. gerrardii	River Current	*riviertaaibos**	NZN
S. glauca	Blue Kuni-Bush	*bloukoenibos*	NZN
S. gracillima var. *glaberrima*	Needle-Leaf Dwarf Currant	*naaldblaar-dwergtaaibos*	NZN
S. grandidens	Sharp-Tooth Currant	*bosnanabessie*	NZN
S. gueinzii	Thorny Karee	*doringkaree*	**inhlokoshiyane,** **umphondo**
S. harveyi	Harveys [*sic*] Currant	*Harvey-se-taaibos*	NZN
S. krebsiana	Mountain or Sour Currant	*suurtaaibos*	**inhlokoshiyane**

Searsia and vernacular names (*cont.*).

Scientific name	English name(s)	Afrikaans name(s)	Zulu name(s)
S. kwazuluana	Kwazulu Dwarf Currant	*kwazulu-dwerg taaibos*	NZN
S. laevigata var. *laevigata*	Dune Currant	*duinetaaibos*	NZN
S. leptocictya var. *leptodictya*	Mountain Karee	*bergkaree*	NZN
S. lucida forma *lucida*	Glossy Currant	*blinktaaibos*	**inhlokoshiyane**
S. montana	Drakensberg Karee	*drakensbergkaree*	**inhlokoshiyane**
S. natalensis	Northern Dune Currant	*noordelike duinetaaibos*	**inhlokoshiyane**
S. nebulosa forma *nebulosa*	Coast Currant	*kustaaibos*	NZN
S. pallens	Ribbed Currant	*oostelike koeniebos bleekkoeniebos*	NZN
S. pentheri	Crow-Berry	*gewone kraaibos*	**inhlokoshiyane**
S. pondoensis	Many-Veined Currant	*veelnerftaaibos*	NZN
S. pterota	Winged Currant	*vlerksteeltaaibos*	NZN
S. pyroides var. *gracilis*	River Firethorn Currant	*riviertaaibos**	**inhlokoshiyane**
var. *integrifolia*	Mountain Firethorn Currant	*bergtaaibos*	**inhlokoshiyane**
var. *pyroides*	Common Wild or Firethorn Currant	*taaibos*	**inhlokoshiyane**
S. refracta	Thorny or Rough-Leaved Currant	*doringkraaibessie*	NZN
S. rehmanniana var. *rehmanniana*	Blunt-Leaf Crow-Berry	*stompblaartaaibos*	**inhlokoshiyane**
S. rigida var. *margaretae*	Margarets [*sic*] Rock Currant	*Margaret-se-kliptaaibos*	NZN
var. *dentata*	Waterberg Currant	*waterbergtaaibos*	NZN
S. rogersii	Rogers [*sic*] Currant	*Rogers-se-taaibos*	NZN
S. rudatisii	Rudatis [*sic*] Dwarf Currant	*Rudatis-se-dwergtaaibos*	NZN
S. tomentosa	Bicolour Currant	*korentebos*	**inhlokoshiyane**
S. transvaalensis	Escarpment Currant	*platorandkaree*	**inhlokoshiyane**
S. tumulicola var. *tumulicola*	Hard-Leaf Currant	*hardeblaartaaibos*	**inhlokoshiyane**

When we look more carefully at the English colloquial names, a distinct pattern or system emerges. The word 'currant' is used in no less than 30 of the 40 English names given (75%). Clearly it is even more an English equivalent of 'Searsia' than 'inhlokoshiyane'. The word 'berry' is used in five names, 'karee' in three names and 'kuni-bush' in two names. Obviously, if the word 'currant' is to be used in the naming of 20 different species of plants, it need to be qualified and we see that this happens on three levels: a primary level, where one of four basic terms (currant, berry, karee, kuni-bush) is chosen; a secondary distinguishing level where one qualifier is added to currants; and a tertiary level where another qualifier is added to the first qualifier. It may be useful to see this in the form of a table ('tree diagram'):

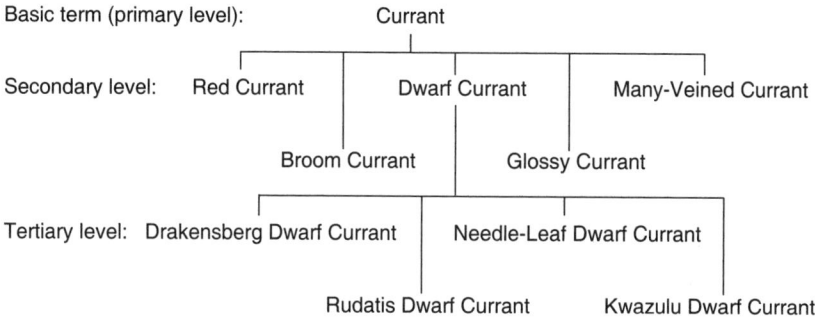

A similar, if simpler, chart can be drawn for 'berry':

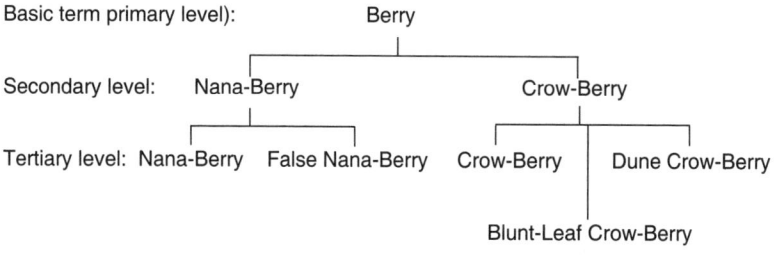

A separate analysis of Afrikaans colloquial names does not have to be done because in almost every case the Afrikaans name is a translation of the English, using '*taaibos*' for 'currant', '*bessie*' for 'berry', '*karee*' for 'karee'[9] and '*koeniebos*' for 'kuni-bush'.

The system of Zulu colloquial names for *Searsia* trees is remarkably similar to the English system; i.e., a choice of basic terms with one (**inhlokoshiyane**) noticeably dominant and then a series of qualifiers at another level. This system is not immediately obvious from the Zulu nomenclature provided by Boon (2010). We need to return to Doke and Vilakazi (1958), who provide a distinctly more complex system to name *Searsia*. To start with, we should note that Boon's names **ikhathabane**, **umdwendwelengcuba** and **umphondo** are not recorded in Doke and Vilakazi, although **umhlabamvubu** is, and it corresponds with Boon's *Searsia* species. Doke and Vilakazi (1958: 313) also give **umhlakothi** for **Rhus villosa* and **Rhus fraseri* and say (1958: 320) that **ihlangushiyane** is used generally for *Rhus* (i.e., *Searsia*) species and that the abbreviated form **inhlokoshane** is often used in place of **inhlokoshiyane** (1958: 332). For the entry **inhlokoshiyane** itself, Doke and Vilakazi say "Small trees of the species [*sic*] Rhus, some with edible berries" (1958: 322) and they then distinguish between:

- **inhlokoshiyane enkulu** (big): *Rhus Legati* [= *legatii*];
- **inhlokoshiyane encane** (little): small *Rhus* shrub;
- **inhlokoshiyane yehlathi** (of the forest): Forest *Rhus*;
- **inhlokoshiyane yehlanze** (of the open veldt): Veld *Rhus*;
- **inhlokoshiyane enoboya** (which has fur): *Rhus villosus* [= *villosa*];
- **inhlokoshiyane emhlophe** (white): *Rhus sonderi*; and
- **inhlokoshiyane yotshani** (of the grass): *Rhus discolor*.

These are clearly the equivalent of the 'secondary level' terms like 'red currant', 'broom currant' and 'dwarf currant' given earlier.

From this, it seems not unreasonable to suggest that the colloquial or vernacular English, Afrikaans and Zulu names are equally systematic, with basic terms like 'currant', 'taaibos' and 'inhlokoshiyane' acting far more like genus names that have to be further qualified by a specific epithet in order to distinguish between species. There is, however, a lack of conformity in the way these 'basic terms' have been treated. It seems reasonably clear that the qualification and further qualification of terms like 'currant' and 'berry (in the

---

9. I would think that the English name is adopted from the Afrikaans one here.

case of *Searsia* English vernacular names) has been a planned and deliberate exercise. In some time past, an individual (or more likely a group of individuals) has taken the conscious decision that 'We need to have a separate and distinct English vernacular name for every plant species'[10] and has then sat down and decided which qualifications and how many layers of these qualifications should be added to basic terms like 'currant', 'berry', 'karee', etc. to get to the required number of distinct colloquial names.

Clearly this planned and deliberate process has not taken place for Zulu plant names. I am in two minds as to whether such a process should or should not take place.

Conformity in Zulu bird names: An experiment

To put this question into a slightly different context, let us look at names for birds in the South African Bantu languages and what is currently taking place with Zulu bird names. Maclean's edition of *Roberts' Birds* (1985) is notable not only for the number of local languages covered, but also because it is the only publication of its kind that explains the motivation behind provision of these names and the way it was done. It is worthwhile quoting the following in full, as so many of the points he makes are valid for Zulu plant names. MacLean explains:

> Bird names in the African languages present far more problems than in the European-derived languages. Many of them are generic (*i.e.* all species of sparrows may have the same name), others are regionally limited in application, one name may be applied to two or more different birds, some well known birds may have more than one name in a different language, and so on. Most bird species have no African names at all. It was therefore decided to scrap all the existing African bird names in past editions of *Roberts'* (many of which were obviously incorrect, or could not be traced to any known language) and to start afresh by soliciting help from experts in the different language groups. They were asked simply to submit as complete and authoritative a list as could be managed in the time available, to make no guesses, and to ensure that the spelling was correct and up-to-date. Considering the

---

10. Or mammal, bird, reptile, fish, insect, etc.

number of African languages in Southern Africa, it is gratifying that so many experts could be found to deal with so many of them . . . In most cases I accepted the linguistic experts' opinions without question, but in others I was able to discuss certain points with them, or at least to query a few individual names that did not seem appropriate. Only time will tell with what degree of co-operative accuracy we have succeeded (1985: xxix).

Van Wyk and Gericke (2000: 144) say that "the power of mythology in perpetuating oral traditional knowledge is vividly demonstrated in the story of 'the shadow of the black-shouldered kite'". The 'story' links fever with the behaviour of birds and shows how the application of 'muthi' made with the plant *Dicoma anomala* will immediately stop a feverish baby behaving like a black-shouldered kite. Van Wyk and Gericke do not identify the culture responsible for this 'mythology', so on reading their account I immediately checked my 1985 *Roberts'* for the Zulu name for this bird to see if the name could shed any light on the issue. Alas! – this widely-distributed and common bird has no Zulu name recorded, although MacLean gives names in Kwangali, Shona, Tsonga, Southern Sotho, Tswana and Xhosa.

There are many more gaps where a bird common and resident in the Zulu-speaking areas has no name recorded in the 1985 edition of *Roberts'*. Among them are the Forest Buzzard (*Buteo oreophilus*), the Steppe Buzzard (*Buteo buteo*), the Palmnut Vulture (*Gypohierax angolensis*), the Black Widowfinch (*Vidua funerea*), the Chorister Robin (*Cossypha dichroa*), the Natal Robin (*Cossypha natalensis*),[11] the Brown-Throated Martin (*Riparia paludicola*) and the Crested Barbet (*Trachyphonus vailantii*).[12] There are many, many more.

As MacLean points out in the quotation above, there are many instances where one name may be applied to two or more birds as with the Black Eagle (*Aquila verreauxii*), the Blackbreasted Snake Eagle (*Circaetus gallicus*), the Martial Eagle (*Polemaetus bellicosus*) and the Crowned Eagle (*Stephanoaetus coronatus*), all of which have the Zulu name *ukhozi*. The last two also share the name *isihuhwa*.

---

11. Although the Cape Robin (*Cossypha caffra*) has two Zulu names: *umbhekle* and *ugaga*.
12. Although Doke and Vilakazi give "*imvunduna*: the Crested Barbet (*Trachyphonus vailantii*)" (1958: 844). Somehow this never made it into the 1985 *Roberts'*.

It was these gaps and duplications in Zulu bird naming that led Noleen Turner[13] to run an initial two-day workshop in September 2013 to explore the possibilities of filling the gaps and reducing duplications. I was one of the participants, as were five Zulu-speaking bird experts – game rangers and birding guides from various game reserves in KwaZulu-Natal. They were quick to pick up the idea of coining new words for 'birds without names', basing their suggestions on existing words for birds in dictionaries that were not linked to identifiable species and otherwise using metaphor, onomatopoeia and other linguistic devices to coin new names. The addition of various suffixes and prefixes to existing names used, as MacLean points out, more as generic names than species names, helped to give distinct names to distinct species in a single genus. The initial workshop was considered a success and another was held over two days in late October 2014.

Some idea of the type of names coined can be seen in the following: Tern: *unolwandle* (< *no-* + *ulwandle* (sea)); Pochard: *isankawu* (something like a vervet monkey, from its call); Pied Barbet: *umhloshana* (the whitish one < *-mhlophe* (white) + diminutive *-ana*); Natal Robin: *unonkositini* (< *inkositini* (concertina, in reference to its melodious and highly variable call); Gannet: *isicibamanzi* (< *isiciba* (arrow) + *amanzi* (water), from the way the bird plunges into the sea). My notes tell me that the name *udemezane* was recorded for the Black-Shouldered Kite, but I did not record the motivation. This was probably an older name, unrecorded in any of the literature but remembered by one of the Zulu-speaking experts.

From Zulu bird names, we move to Zulu insect names. From 2011 to 2012, Jessica Cockburn, a postgraduate student in Rhodes University's department of zoology and entomology, conducted fieldwork on Zulu insect names. A total of 67 people were interviewed in eleven different localities in KwaZulu-Natal. The results of her research are given in Cockburn, Khumalo-Seegelken and Villet (2014). From the information in this article, it is clear that indigenous nomenclature for insects shares the same features as those for birds and plants: not every species has its own name, some names are used as generic names by two or more species and one insect may have several names because of regional differences. Cockburn's research shows that for a number of insects there is a clear regional distribution of names. For example, north-east of the uThukela River a butterfly is commonly known as *isiphaphalazi* and south-east of the

---

13. Associate Professor in Zulu Studies in the School of Arts, University of KwaZulu-Natal.

river more commonly known as *uvemvane*. Cockburn, Khumalo-Seegelken and Villet have some interesting comments to make about attempts to reconcile 'folk taxonomies' with scientific systems of biological nomenclature, and on attempts to standardise the names in an indigenous naming system. We shall look at these comments in the next section.

Where to next with Zulu plant names?

In the context of the various problems with vernacular naming in the natural sciences, let us look again at Zulu plant names. It is abundantly clear that indigenous botanical nomenclature faces the same issues that MacLean described for the indigenous names for birds. Substitute 'lilies', say, for 'sparrows', and 'plants' for 'birds' and MacLean has described the situation perfectly:

> Many of them are generic (*i.e.* all species of sparrows may have the same name), others are regionally limited in application, one name may be applied to two or more different birds, some well known birds may have more than one name in a different language, and so on. Most bird species have no African names at all (1985: xxix).

The question remains: what to do about it? The way I see it, there are three things that can be done:

- Nothing: leave the problem as it is. Acknowledge that the situation as described by MacLean is a feature of any indigenous knowledge system. Argue that the problems are only caused by attempts to shape an indigenous knowledge system to conform to Linnaean-type taxonomies and classification. Argue that in South Africa it is important for African indigenous knowledge systems to remain African and not be reshaped to suit Western concepts of naming and identification. Cockburn, Khumalo-Seegelken and Villet are very clear on this point:

> Attempts to compare non-scientific naming systems (often termed 'folk taxonomies') such as those found in isiZulu or English with any terminological systems of biological nomenclature have ultimately been rejected as ineffective and undesirable. This situation is especially true when it comes to trying to promote and support inter-related networks of

knowledge systems within a given context in post-colonial Africa. Speakers of an indigenous African language such as isiZulu perceive and experience such attempts as biased, presumptuous and exclusively Eurocentric (2014: 7).

- <u>Intensive fieldwork and research</u>: acknowledge the indigenous nature of the existing Zulu botanical nomenclature and attempt to record as much oral knowledge as possible. Correct all errors in existing publications. Do intensive fieldwork to establish which of many names for one plant are regional variations, which are names used only by specialists such as traditional healers, and so on. Record this information with as much information as possible on plant usage in Zulu societies. Publish this information in comprehensive dictionaries and other reference books. Recognise that the indigenous knowledge system is still a <u>system</u> and attempt to recognise the nature of the system, such as seeing the name *inhlokoshiyane* as a genus rather than a specific name that just happens to be used for 22 species of plant. Basically, these were the goals of the ZBKP. To achieve them requires massive financial input, from government and other sources. It was the lack of financial resources that led to the gradual death of the ZBKP in the mid-2000s. This was the approach adopted by Cockburn in her research into Zulu insect names (Cockburn, Khumalo-Seegelken and Villet 2014).

- <u>Conformity workshops</u>: hold workshops of the sort organised for Zulu bird names by Noleen Turner in September 2013 and October 2014 with Zulu-speaking botanists and other plant experts as well as Zulu linguists, and attempt to do what has been done with English and Afrikaans in the natural sciences books mentioned above. 'Fill in the gaps' and ensure that every species of plant found in the Zulu-speaking area has its own individual name, even if this ends up as an equivalent of something like 'Drakensberg Dwarf Currant' (*drakensbergdwergtaaibos* in Afrikaans). Such a procedure may mean accepting a name like **inhlokoshiyane** as a genus name; qualifying it, say, with *enkulu* (big), using **inhlokoshiyane enkulu** as a name for four or five similar species; and then further qualifying them as, say, **inhlokoshiyane enkulu yehlathi** (big forest inhlokoshiyane) and **inhlokoshiyane enkulu enameva** (big thorny inhlokoshiyane). This would be a tedious and time-consuming project, but fairly mechanical. It would require some financial outlay, but not nearly as much as the extensive fieldwork outline in the above suggestion. Some might say that adopting this process would be good in that the system of Zulu plant naming would conform to the English and Afrikaans names and even, to some extent, to the system of scientific nomenclature. Others might say that such an approach is bad for the very same reason: that an African indigenous knowledge system has lost its traditional identity by being made to conform to a Western-type knowledge system. In effect, this is a process of standardisation and Cockburn, Khumalo-Seegelken and Villet have some useful comments to make about it:

> Linguistic standardization is part of the mandate of the Pan South African Language Board, which must reconcile its aims with all . . . aspects of diversity and evolution in isiZulu entomological names, which also occur in isiZulu ornithological names, isiZulu botanical names . . . folk taxonomies worldwide and language in general (2014: 9).

They go on to make pertinent comments about the role of dictionaries in standardisation processes:

> Traditionally, dictionaries have been seen by their users (and often by their writers) as arbiters of standardized usage and spelling, thus accorded a prescriptive role. Modern lexicography recognizes that languages are dynamic and evolutionary, an insight that has led to dictionaries becoming linguistically descriptive, rather than prescriptive, but they can still be compiled for linguistic standardization agendas (2014: 9).

The aims of this book

In this book I have not attempted to 'fix' any 'problems' with the Zulu botanical naming system. My approach has been linguistic and onomastic, and certainly descriptive rather than prescriptive. I have focused on the Zulu names of plants, not on the plants themselves. I have attempted to describe purely linguistic strategies in the formation of names: derivational process where verbs become nouns, and nouns spawn other nouns. I have looked at processes of prefixing, suffixing, reduplication of stems, and processes of compounding. I have looked at literary aspects, such as the use of metaphor, in the creation of plant names.

Above all, I have tried to show that just as a language is an integral part of the culture that speaks that language, so also plant names, words in a particular language, are part of the culture. At all stages I have tried to show that in plant names we are looking at an interface between language, culture and the world of plants. We have seen that the name **umkhondo** (Caterpillar Bean, *Zornia capensis*) refers to the way this plant, tied around the ankles of women carrying new-born babies, will prevent them picking up bad luck when crossing the spoor of ill-omened animals. We have seen that travelling with a stalk of **insutha** grass in your hair will ensure you hospitality on the road. We have seen that carrying a sprig of the **iphinganhloya** plant on your head will help to purify you after killing someone in battle. These, and many, many other examples, show how plant, plant name and cultural beliefs are linked.

This book is not a dictionary of Zulu plant names. It is not intended as a reference work for botanists. I have left out hundreds of Zulu plant names, simply because they have not provided any further information on the links between language, culture and plant.

When I started studying Zulu as a young university student in 1970, I saw it as a language of beauty and poetry, as a language of great ingenuity, capable of creating imaginative new words on a regular basis, and as a language providing a deep insight into the culture of the people who speak it. After more than twelve years working specifically with Zulu plant names, I have not changed any of those views.

Bibliography

Ankarloo, Bengt, Clark, Stuart and Monter, William (eds). 2002. *Witchcraft and Magic in Europe: The Period of the Witch Trials*. London: Athlone Press.

Bailey, Liberty H. 1963. *How Plants Get Their Names*. New York: Dover Publications.

Berglund, Axel-Ivar. 1976. *Zulu Thought-Patterns and Symbolism*. Cape Town: David Philip.

Berlin, B. 1976. 'The Concept of Rank in Ethnobiological Classification: Some Evidence from Aguarana Folk Botany'. *American Ethnologist* 3: 381–99.

Berlin, B., Breedlove, D.E. and Raven, P.H. 1973. 'General Principles of Classification and Nomenclature in Folk Biology'. *American Anthropologist* 75: 215–42.

Bews, John. 1921. *An Introduction to the Flora of Natal and Zululand*. Pietermaritzburg: City Printing Works.

Blessing, Henrik Greve *see* Paulsen, Berit S. et al.

Boon, Richard. 2010. *Pooley's Trees of Eastern South Africa*. Durban: Flora and Fauna Publications Trust.

Botha, Theunis J.R. 1977. *Watername in Natal: 'n Inleiding tot die Studie van Zoeloeplekname*. Pretoria: Raad vir Geesteswetenskaplike Navorsing.

Bristow, Alec. 1979. *The Sex Life of Plants: A Study of the Secrets of Reproduction*. London: Barrie & Jenkins.

Bryant, Alfred T. 1905. *A Zulu-English Dictionary with Notes on Pronunciation . . .* Pinetown: Mariannhill Mission Press.

———. 1909. 'Zulu Medicine and Medicine Men'. *Annals of the Natal Museum* II(1): 1–103.

———. 1967. *The Zulu People As They Were Before the White Man Came*. Pietermaritzburg: Shuter and Shooter.

———. 1970. *Zulu Medicine and Medicine-Men*. Cape Town: C. Struik.

Bryson, Bill. 1994. *Made in America: An Informal History of the English Language in the United States*. London: Black Swan.

Callaway, Henry. 1970. *The Religious System of the AmaZulu*. Cape Town: C. Struik.

Carroll, Lewis. 1871. *Through the Looking-Glass and What Alice Found There*. London: Macmillan.

Clinning, Charles. 1989. *Southern African Bird Names Explained*. Johannesburg: Southern African Ornithological Society.

Coates Palgrave, Keith. 1977. *Trees of Southern Africa*. Cape Town: C. Struik.

Coates Palgrave, Meg. 2002. *Palgrave's Trees of Southern Africa*. Cape Town: C. Struik.

Cockburn, J.J., Khumalo-Seegelken, B. and Villet, M.H. 2014. 'IziNambuzane: IsiZulu names for Insects'. *South African Journal of Science* 110(9&10): 1–13.

Colenso, John W. 1878. *Zulu-English Dictionary*. Pietermaritzburg: P. Davis.

Collins Dictionary of the English Language. 1986. London: Collins.

Cunningham, Anthony B. 1985. 'The Resource Value of Indigenous Plants to Rural People in a Low Agricultural Potential Area'. PhD thesis, University of Cape Town.

Doke, Clement and Vilakazi, B.W. 1958. *Zulu-English Dictionary*. Johannesburg: Witwatersrand University Press.

Frazer, Sir James George. 2002. *The Golden Bough*. New York: Dover Publications.

Gardner, Martin. 1960. *The Annotated Alice*. London: Penguin.

Gerstner, Jacob. 1938. 'A Preliminary Check List of Zulu Names of Plants with Short Notes'. *Bantu Studies* 12: 215–36 and 321–42.

———. 1939. 'A Preliminary Check List of Zulu Names of Plants with Short Notes'. *Bantu Studies* 13: 49–64, 131–49 and 307–26.

———. 1941. 'A Preliminary Check List of Zulu Names of Plants with Short Notes'. *Bantu Studies* 15: 277–301 and 369–83.

Guest, Bill. 2010. *A Fine Band of Farmers Are We: A History of Agricultural Studies in Pietermaritzburg 1934–2009*. Pietermaritzburg: Natal Society Foundation.

Hammond-Tooke, W.D. 1962. *Bhaca Society: A People of the Transkeian Uplands, South Africa*. Cape Town: Oxford University Press.

Hanks, Patrick (ed.) *see Collins Dictionary of the English Language*.

Hedges, David W. 1978. 'Trade and Politics in Southern Mozambique and Zululand in the Eighteenth and Early Nineteenth Centuries'. PhD thesis, University of London.

Heiser, Charles B. Jr. 1973. *Seed to Civilisation: The Story of Man's Food*. San Francisco: W.H. Freeman.

———. 1985. *Of Plants and People*. Norman: University of Oklahoma Press.

Helander-Renvall, Elina M. 2007. 'Traditional Ecological Knowledge, Snow and Sámi Reindeer Herding' in *Knowledge and Power in the Arctic: Conference Proceedings*, edited by P. Kankaanpää. Rovaniemi: University of Lapland: 87–99.

Hutchings, Anne. 1996. *Zulu Medicinal Plants: An Inventory*. Pietermaritzburg: University of Natal Press.

Koopman, Adrian. 1983. 'Zulu Place-Names in the Drakensberg' in *G.S. Nienaber: 'n Huldeblyk*, edited by A.J.L. Sinclair. Cape Town: University of the Western Cape: 297–306.

———. 2002. *Zulu Names*. Pietermaritzburg: University of Natal Press.

———. 2005. 'Unpacking Jamludi: An Exercise in Interdisciplinary Onomastics'. *Nomina Africana* 19(2): 159–84.

————. 2009. 'A Re-Look at the Semantics of "eThekwini", or Why Durban's Mayor Needs More Milk in his Tea'. *Nomina Africana* 23(2): 33–60.

————. 2011. 'Lightning Birds and Thunder Trees'. *Natalia* 41: 40–60.

————. 2013(a). 'The Interface between Magic, Plants, and Language'. *Southern African Humanities* 25: 61–77.

————. 2013(b). 'Pandas in the Forests of KwaZulu-Natal?' *Natalia* 43: 150–2.

————. 2013(c). 'Prepuce Covers and the Ikhamanga'. *Natalia* 43: 152–5.

————. 2014. 'Henrik Greve Blessing's *South African Traditional Medicinal Plants from KwaZulu-Natal*' (book review). *Natalia* 44: 94–8.

Krige, Eileen Jensen. 1950. *The Social System of the Zulus*. Pietermaritzburg: Shuter and Shooter.

Louwrens, L.J. 1999. 'An Ethnobiological Investigation into Northern Sotho Plant Names' in *African Mosaic: A Festschrift for JA Louw*, edited by Rosalie Finlayson. Pretoria: UNISA Press: 285–310.

Mabille, Adolphe and Dieterlen, H. 1950. *Southern Sotho-English Dictionary*. Reclassified, revised and enlarged by R.A. Paroz. Morija: Morija Sesuto Book Depot.

MacLean, Gordon. 1985. *Roberts' Birds of Southern Africa*. Fifth edition. Cape Town: John Voelcker Bird Book Fund.

McLachlan, G.R. and Liversidge, R. 1978. *Roberts Birds of South Africa*. Cape Town: John Voelcker Bird Book Fund.

McLaughlin, Terence. 1971. *Coprophilia or a Peck of Dirt*. London: Cassell.

Moll, Eugene. 1981. *Trees of Natal: A Comprehensive Field Guide to Over Seven Hundred Indigenous and Naturalized Species*. Cape Town: University of Cape Town Eco-Lab Trust Fund.

Ngwenya, Mkhipheni, Koopman, Adrian and Williams, Rosemary. 2003. *Ulwazi LwamaZulu Ngezimila: Isingeniso/Zulu Botanical Knowledge: An Introduction*. Durban: National Botanical Institute.

Nicolaisen, W.F.H. 1974. 'Names as Verbal Icons'. *Names* 22(3): 104–10.

Ong, Walter J. 1982. *Orality and Literacy: The Technologizing of the Word*. London: Methuen.

Palgrave *see* Coates Palgrave.

Palmer, Eve and Norah Pitman. 1972. *Trees of Southern Africa Covering All Known Indigenous Species in the Republic of South Africa, South-West Africa, Botswana, Lesotho and Swaziland* (3 volumes). Cape Town: A.A. Balkema.

Patterson, Rod. 1987. *Reptiles of Southern Africa*. Cape Town: Struik.

Paulsen, Berit S., Ekeli, H., Johnson, Q. and Norum, K.R. (eds). 2012. *Dr. Henrik Greve Blessing: South African Medicinal Plants from KwaZulu-Natal: Described 1903–1904*. Norway: Unipub.

Pavord, Anna. 2005. *The Naming of Names: The Search for Order in the World of Plants*. London: Bloomsbury.

Pooley, Elsa. 1993. *The Complete Field Guide to Trees of Natal, Zululand and Transkei.* Durban: Natal Flora Publications Trust.

———. 1998. *A Field Guide to Wild Flowers: KwaZulu-Natal and the Eastern Region.* Durban: Natal Flora Publications Trust.

———. 2003. *Mountain Flowers: A Field Guide to the Flora of the Drakensberg and Lesotho.* Durban: Flora Publications Trust.

Roberts, Margaret. 1990. *Indigenous Healing Plants.* Johannesburg: Southern Book Publishers.

Rycroft, David K. and Ngcobo, A.B. 1988. *The Praises of Dingana: Izibongo zikaDingana.* Pietermaritzburg: University of Natal Press.

Rymer, Russ. 2012. 'Vanishing Voices'. *National Geographic* 222(1): 60–93.

Saarelma, M.M. 2009. 'African and European Anthroponymic Systems: Similarities and Differences'. *Onoma* 44: 191–217.

Samuelson, R.C.A. 1923. *The King Cetywayo Zulu Dictionary.* Durban: Commercial Printing Co.

Schultes, Richard E. and Hofmann, A. 1992. *Plants of the Gods: Their Sacred, Healing and Hallucinogenic Powers.* Rochester: Healing Arts Press.

Skaife, S.H. 1979. *African Insect Life.* Cape Town: Struik.

Smith, Christo A. 1966. *Common Names of South African Plants.* Pretoria: Government Printer.

Smithers, Reay H.N. 1983. *The Mammals of the Southern African Subregion.* Pretoria: University of Pretoria.

Stevens, John E. 2008. *Discovering Wild Plant Names.* Aylesbury: Shire Publications.

Stirton, Charles H. (ed.). 1978. *Plant Invaders: Beautiful but Dangerous.* Cape Town: Department of Nature and Environmental Conservation of the Cape Provincial Administration.

Stratton, Kimberly B. 2007. *Naming the Witch: Magic, Ideology and Stereotype in the Ancient World.* New York: Columbia University Press.

Thurber, James. 1962. *Thurber Country.* London: Penguin.

Titchmarsh, Alan. 2008. 'Heaven in a Wild Flower' in *Icons of England*, edited by Bill Bryson. London: Transworld Publishers: 324–7.

Turner, Victor. 1977. *The Ritual Process: Structure and Anti-Structure.* New York: Cornell University Press.

Twain, Mark. [1897] 1989. *Following the Equator: A Journey Around the World.* Toronto: General Publishing Company.

Van Warmelo, Nicolaas J. 1976. 'Who are the Basotho?' *African Studies* 35(2): 91–8.

Van Wyk, Ben-Erik and Gericke, Nigel. 2000. *People's Plants: A Guide to Useful Plants of Southern Africa.* Pretoria: Briza Publications.

Van Wyk, Braam, Van den Berg, Erika, Coates Palgrave, Meg and Jordaan, Marie. 2011. *Dictionary of Names for Southern African Trees: Scientific Names of Indigenous*

Trees, Shrubs and Climbers with Common Names from 30 Languages. Pretoria: Briza Publications.

Wagner, Vincent A. 1986. *Frogs of South Africa*. Johannesburg: Delta Books.

Walker, Clive. 1981. *Signs of the Wild*. Johannesburg: Natural History Publication.

Watt, John Mitchell and Breyer-Brandwijk, Maria Gerdina. 1962. *The Medicinal and Poisonous Plants of Southern and Eastern Africa*. Second edition. London: E. & S. Livingstone.

Webb, Colin de B. and Wright, J.B. (eds). 1976–2014. *The James Stuart Archive of Recorded Oral Evidence Relating to the History of the Zulu and Neighbouring Peoples* (6 volumes). Pietermaritzburg: University of Natal Press and Durban: Killie Campbell Africana Library.

Wild, Hiram. 1952. *A Southern Rhodesian Botanical Dictionary of Native and English Plant Names*. Salisbury: Government Printer.

Winick, Charles. 1964. *Dictionary of Anthropology*. Totavia, NJ: Littlefield, Adams and Company.

Wright, John. 2012. 'A.T. Bryant and the "Lala"'. *Journal of Southern African Studies* 38(2): 355–68.

Zukulu, Sinegugu, Dold, Tony, Abbott, Tony and Raimondo, Domitilla. 2012. *Medicinal and Charm Plants of Pondoland*. Pretoria: South African National Biodiversity Institute.

Glossary

This glossary contains both botanical (bot.) and linguistic (ling.) terms.

affix (ling.) = any grammatical morpheme whether a **suffix, prefix** or infix. In 'unbending', 'un-' and '-ing' are affixes.

anthroponym (ling.) = the name of a person. Related terms are first name, personal name, given name, baptismal name, Christian name, surname, clan name, family name, nickname and praise name (> anthroponymy, anthroponomastics, anthroponymical system).

appellative (ling.) = any noun that is not a proper name. 'Dog', 'hound', 'terrier' and 'spaniel' are appellatives; 'Rover', 'Prince', 'Blackie' and 'Spot' are proper names. Also as common noun.

binomial (bot.) = a name consisting of two words. Names of species comprise the genus name plus the species name (or epithet), e.g. *Calpurnia aurea*. Scientific names of all living organisms are written in the Latin form and comply with the rules governing Latin grammar, including words derived from other languages that are also treated as Latin in form. Taxonomic ranks below the level of species, as for example subspecies and varieties are called trinomial names because they comprise three words, namely genus + species + subspecies, e.g., *Vigna unguiculata* subsp. *stenophylla*; or genus + species + variety, e.g., *Acacia luederitzii* var. *luederitzii*.

botonomasticon (ling.): see **onomasticon**.

complex (ling.) = usually used of a **noun stem** that can be broken down into smaller meaningful units, of which only one is a **root**. For example, in the plant name **isimiselo**.

compound (ling.) = used of a noun stem containing two or more parts that can stand on their own as separate words.

dendronomasticon (ling.): see **onomasticon**.

derivation (ling.) = the creation of new lexemes from an existing one through the addition of derivational affixes. Zulu is a highly derivational language creating nouns from verbs (*inkalipho* (cleverness) < *khalipha* (be clever)), nouns from noun (*inkomazi* (cow) < *inkomo* (head of cattle)) and *uSipho* (Mr Gift) < *isipho* (gift)) and so on. Many of the plant names in this book are derived from (<) other parts of speech (> derivational process, derivational affix).

epiphyte (bot.) = a plant that grows on another plant without deriving nutrition from it, e.g., *Ansellia africana*.

family (bot.) = a higher taxonomic unit or level of classification containing one genus or two or more genera.

family (ling.) = a group of languages understood to be derived historically from a 'mother language'.

genera (bot.) = plural of **genus**.

genus (bot.) = taxonomic unit or level of classification comprising one or more species.

grammar (ling.) = the rules that govern the operation of words in a language, including such aspects as forming different tenses of a verb, pluralising a noun, turning a positive into a negative, etc.

grammatical morpheme (ling.) = an **affix** used to execute grammatical rules.

habitat (bot.) = the total of biotic (living) and abiotic (non-living) elements that comprise the environment of the plant.

lexeme (ling.) = a word as a meaningful unit.

lexical (ling.) = referring to 'real life' meaning. In the word 'jumping', 'jump' has the lexical 'real' meaning 'leap upwards or over something'; '-ing' is a grammatical affix. It is 'jump' that is recorded in a **lexicon**, not 'jumping'.

lexicon (ling.) = in any language, a list of words and their meanings. All of us carry at least one lexicon in our heads. A published lexicon is usually called a dictionary.

morpheme (ling.) = a minimal unit of meaning in a language, which cannot be broken down into smaller units. A word like 'jump' has only one morpheme, 'jumping' has two. A word like 'antidisestablishmentarianism' has several.

morphology (bot.) = the outward appearance or features of the 'body' of the plant. This term usually applies to characteristics that can be seen with the unaided eye, for example the overall shape or growth form (habit) of the plant and features of its leaves, flowers, fruits and bark.

morphology (ling.) = the study of word structure, typically looking at word roots and various **affixes**.

noun class (ling.) = normally used of the Bantu languages, a noun class is a group of nouns characterised by a common prefix ('noun class prefix') and the plural is normally another specific noun class. In Zulu, *umu-ntu* (person) is in noun class 1, with the plural *aba-ntu* in noun class 2. The noun *um-hlanga* (reed) is in noun class 3, with its plural *imi-hlanga* in noun class 4. *-Ntu* and *-hlanga* are lexical morphemes; *umu-, aba-, um-* and *imi-* are grammatical morphemes.

onomasticon (ling.) = a combination of the words 'onomastic' and 'lexicon', referring to a list of names, often with their derivation and meaning. Note also anthroponomasticon, a list of personal names; toponomasticon, a list of place names; botonomasticon a list of plant names; and dendronomasticon, a list of tree names.

onomastics (ling.) = the study of proper names, naming and naming systems > onomastician, **onomasticon**.

phoneme (ling.) = a unit of sound in speech, not always accurately represented by letters. For example, the word 'thought' has seven letters, but only three phonemes.

phonological process (ling.) = the changing of phonemes in a word as a result of derivational processes. For example, the noun stem *-khosi* (in *amakhosi* (chiefs)) becomes *-kosi* in the singular; the verb root *bamb-* becomes *banj-* in the passive (*ngiyambamba* (I catch him) versus *Ngibanjwa nguye* (I am caught by him)); and the noun stem *-bob(o)* becomes *-botsh-* in the locative (*imbobo* (hole) versus *embotshweni* (in the hole)). Many of the plant names in this book have undergone phonological processes as they were derived from other parts of speech.

phonology (ling.) = the study of the speech sounds in a particular language.

prickle (bot.) = usually short, sharply pointed outgrowth from the stem, e.g., *Rubia cordifolia*.

proper name (ling.) = very difficult to define and perhaps easier to exemplify: 'table', 'mountain' and 'bay' are appellatives; 'Table Mountain' and 'Table Bay' are proper names. The words 'daisy', 'iris' and 'rose' are appellatives when applied to plants; but when applied to girls they are the proper names 'Daisy', 'Iris' and 'Rose'.

reduplication (ling.) = the repetition of a root morpheme in a word for a certain effect, e.g., Zulu *imisindomsindo* (confusing babble of different sounds) < *imisindo* (sounds); *bukabuka* (glance quickly at) < *buka* (look at).

root (bot.) = part of a branching system of a plant, usually underground, that anchors the plant and conducts minerals and water from the soil into the plant.

root (ling.) = the lexical morpheme in any language. In the plant name **isimiselo** (*Gloriosa superba*) the verb *m(a)* (stand, stand up) is the lexical or root morpheme (or simply the 'root'). On either side of it are the prefix *isi-* and the suffixes, *-is-*, *-el-* and *-o*.

samara (bot.) = a fruit with a wing, e.g., *Pterocarpus angolensis*. The wing is an outgrowth of the fruit wall. In mature fruits wings are dry and thin and use wind currents to carry seeds away from the parent plant. This reduces the competition for resources such as light and nutrients between the parent plant and seedling offspring.

saprophyte (bot.) = a plant aided by a fungus that obtains its nutrients from decaying organic matter. Saprophytic plants typically lack leaves and the green pigment chlorophyll.

simple (ling.) = usually used of a **noun stem** that cannot be broken down into smaller meaningful units.

spadix (bot.) = thick, fleshy column-like structure bearing many small and highly modified flowers that do not resemble 'typical' flowers, e.g., in genera *Arum*, *Zantedeschia*.

spathe (bot.) = large, broad bract or modified, often colourful leaf surrounding the spadix, e.g., *Arum*, *Zantedeschia*, or thickened, boat-shaped and green or colourful surrounding cluster of flowers, e.g., *Strelitzia*.

species (bot.) = a collection of individual organisms, recognisably similar to each other and generally capable of sexual reproduction among themselves. Such individuals are usually incapable of producing viable and fertile offspring with individuals of other species.

spine (bot.) = hard, pointed, straight or curved, often needle-like, modified leaf, e.g., *Pereskia aculeata*.

stem (ling.) = usually used of Bantu languages, the part of the noun left when the noun class prefix is removed. Stems may be **simple, complex** or **compound**.

subsp. (bot.) = abbreviation of sub-species.

sub-species (bot.) = within a species this refers usually to a group or population of reproductively isolated individuals that, over time, have inherited one or more characteristics that distinguish them from other populations, i.e., other subspecies.

suffix (ling.) = an **affix** placed after the root of a word. See **root** (ling.).

syn. (bot.) = abbreviation of synonym.

synonym (bot.) = one of two or more names that are given to the same species. The current or valid name is the one that should be used when referring to a given species. A synonym is more specifically a name by which a species was formerly known, but which is no longer used as it is no longer the `correct' name. For example, *Podocarpus falcatus* is the previously accepted or valid name (i.e., now a synonym) of a species of Yellowwood currently more correctly known as *Afrocarpus falcatus*; and *Strychnos gerrardii* is a synonym of the plant now more correctly referred to as *Strychnos madagascariensis*.

synonym (ling.) = one of two or more words with similar meaning, like 'leap' and 'jump', or 'joy' and 'happiness'.

synonymy (bot.) = the existence of two or more names for the same species.

taxonomic unit (bot.) = a level or rank of classification. For example, more or less identical individuals capable of sexual reproduction among themselves constitute a species: different species are placed in the same genus and a number of genera sharing core similarities are placed together in a family.

tendril (bot.) = long, flexible, whip-like tip of a stem or leaf that coils around an external structure and allows the plant to support or elevate itself, e.g., *Gloriosa superba*.

thorn (bot.) = a short or long, straight or curved, usually hard, pointed branch, e.g., species of *Acacia, Dalbergia armata*.

toponym (ling.) = place name, geographical name > 'toponymy', 'toponomastics'.

tree diagram (ling.) = a means of illustrating the structure of a sentence (sometimes a word) with a diagram showing the constituents of a sentence (or words) and the relationship between them. For example, a sentence may be divided between noun phrase and verb phrase at the first level of analysis and these may be broken down further as [preposition + noun + adjective] in the case of the noun phrase and [verb + object + adverb] in the verb phrase.

var. (bot.) = abbreviation of variety.

variety (bot.) = a level of classification within a species or sub-species. It refers to individuals within a population with one or two distinctive features, for example, leaf shape.

v.l. (ling.) = abbreviation for variant, from the Latin *vario lecto*.

List of synonyms

Acalypha villicaulis > *Acalypha brachiata*

Adhatoda Andromeda > *Duvernoia Andromeda*

Agapanthus umbellatus > *Agapanthus africanus*

Albizia mossambicensis Sim. > *Albizia versicolor*

Alchemilla vulgaris > *Alchemilla xanthochlora*

Andropogon sorghum > *Sorghum bicolour*

Anthericum elongatum > *Trachyandra revoluta*

Anthericum humile > *Anthericum angustifolium*

Aptenia cordifolia > *Mesembryanthemum cordifolium*

Asclepias physocarpa > *Gomphocarpus physocarpus*

Asparagus plumosus > *Asparagus setaceus*

Asparagus sprengeri > *Asparagus aethiopicus*

Aster muricatus > *Felicia muricata*

Canthium mundianum > *Afrocanthium mundianum*

Canthium obovatum > *Psydrax obovata*

Cassia occidentalis > *Senna occidentalis*

Cassinia phylicifolia > *Tenrhynea phylicifolia*

Cassipourea gummiflua var. *verticillata* > *Cassipourea gummiflua*

Cassipourea verticillata > *Cassipourea gummiflua*

Chenopodium ambrosioides > *Dysphania ambrosioides*

Chlorocodon whitei > *Mondia whitei*

Cissus connivens > *Cyphostemma natalitium*

Cissus lanigera > *Cyphostemma lanigerum*

Clerodendrum glabrum > *Volkameria glabra*

Coccinia palmata > *Coccinia rehmannii*

Colpoon compressum > *Osyris compressa*

Combretum glomeruliflorum > *Combretum erythrophyllum*

Combretum suluense > *Combretum collinum* subsp. *suluense*

Conopharyngia ventricosa > *Tabernaemontana ventricosa*

Coronopsis didymos > *Lepidium didymium*

Crassula platyphylla > *Crassula nudicaulis* var. *platyphylla*

Crassula turrita > *Crassula capitella* subsp. *thyrsiflora*

Curtisia faginea > *Curtisia dentata*

Cussonia umbellifera > *Schefflera umbellifera*

Dicoma zeyheri > *Macledium zeyheri*

Diospyros nummularia > *Diospyros natalensis* subsp. *Nummularia*

Eclipta erecta > *Eclipta prostrata*

Ekebergia velutina > *Ekebergia benguelensis*

Elephantorrhiza burchellii > *Elephantorrhiza elephantina*

Elionurus argenteus > *Elionurus muticus*

Eriospermum abyssinicum > *Eriospermum flagelliforme*

Euclea daphnoides > *Euclea racemosa* subs. *daphnoides*

Euclea multiflora > *Euclea natalensis* subsp. *angolensis*

Eulophia robusta > *Eulophia livingstoneana*

Euphorbia pugniformis > *Euphorbia procumbens*

Ficus hippopotami > *Ficus trichopoda*

Ficus sonderi > *Ficus glumosa*

Gardenia neuberia > *Hyperacanthus amoenus*

Gasteria glabra > *Gasteria carinata* var. *carinata*

Gnidia kraussii > *Lasiosiphon kraussii*

Grumilea capensis > *Psychotria capensis*

Harpochloa capensis > *Harpochloa falx*

Helichrysum leiopodium > *Helichrysum nudifolium*

Hippobromus alata > *Hippobromus pauciflorus*

Homeria pallida > *Moraea pallida*

Hypoxis latifolia > *Hypoxis colchicifolia*

Kalanchoe thyrsiflora > *Kalanchoe tetraphylla*

Lachnopylis floribunda > *Nuxia floribunda*

Landolphia capensis > *Ancylobothrys capensis*

Lantana rugosa > *Buddleja salviifolia*

Lippia asperifolia > *Phyla scaberrima*

Litanthus pusillus > *Drimia uniflora*

Littonia modesta > *Gloriosa modesta*

Maerua woodii > *Bachmannia woodii*

Maesa rufescens > *Maesa lanceolata* var. *rufescens*

Millettia sutherlandia > *Lonchocarpus sutherlandii*

Moraea spathacea > *Moraea spathulata*

Moraea spathulata > *Moraea pallida*

Moschosma riparia > *Tetradenia riparia*

Myrica serrata > *Morella serrata*

Myrsine melanophloeos > *Raphanea melanophloeos*

Mystroxylon aethiopicum > *Cassine aethiopica*

Oeceoclades mackenii > *Oeceoclades maculata*

Oldenlandia decumbens > *Oldenlandia affinis* subsp. *fugax*

Olea foveolata > *Chionanthus foveolatus*

Olea laurifolia > *Olea capensis*

Oncinotis inandensis > *Oncinotis tenuiloba*

Ophiocaulon gummifera > *Adenia gummifera*

Oxyanthus gerrardii > *Oxyanthus speciosus* subsp. *gerrardii*

Pachystigma pygmaeum > *Vangueria pygmaea*

Pavetta assimilis > *Pavetta gardeniifolia*

Pelargonium aconitophyllum > *Pelargonium luridum*

Penanisia variabilis > *Penanisia prunelloides*

Phytolacca abyssinica > *Phytolacca dodecandra*

Piliostigma thonningii > *Bauhinia thonningii*

Piptadenia buchananii > *Newtonia buchananii*

Plumbago capensis >*Plumbago auriculata*

Podocarpus falcatus >*Afrocarpus falcatus*

Popowia caffra > *Monanthotaxis caffra*

Protea hirta > *Protea welwitschii*

Pygeum africanum > *Prunus africana*

Ranunculus pinnatus > *Ranunculus wallichianus*

Rauvolfia natalensis > *Rauvolfia caffra*

Rhynchelytrum setifolium > *Melinis nerviglumis*

Royena lucida > *Diospyros whyteana*

Sapium ellipticum > *Schirakiopsis elliptica*

Sapium integerrimum > *Sclerocroton integerrimus*

Sapium reticulatum > *Sclerocroton integerrimus*

Schizoglossum woodii > *Aspidoglossum woodii*

Schotia transvaalensis > *Schotia capitata*

Schrebera mazoensis > *Schrebera alata*

Schrebera saundersiae > *Schrebera alata*

Scilla nervosa > *Merwilla nervosa*

Sclerocarya caffra > *Sclerocarya birrea*

Senecio dieterlenii > *Senecio rhomboideus*

Sonchus ecklonianus > *Sonchus dregeanus*

Stangeria paradoxa > *Stangeria eriopus*

Strelitzia augusta > *Strelitzia alba*

Strychnos gerrardii > *Strychnos madagascariensis*

Synadenium arborescens > *Euphorbia cupularis*

Synadenium cupulare > *Euphorbia cupularis*

Tapinanthus natalitius > *Agelanthus natalitius*

Tecomaria capensis > *Tecoma capensis*

Testudinaria elephantipes > *Dioscorea elephantipes*

Trema bracteolata > *Trema orientalis*

Trichopteryx simplex > *Loudetia simplex*

Urginea delagoensis > *Drimia delagoensis*

Vellozia retinervis > *Xerophyta retinervis*

Vernonia anisochaetoides > *Distephanus anisochaetoides*

Vernonia hirsuta > *Hilliardiella hirsuta*

Vernonia podocoma > *Gymnanthemum myrianthum*

Vigna glabra > *Vigna luteola*

Vigna triloba >*Vigna unguiculata* subsp. *stenophylla*

Vitex wilmsii > *Vitex obovata*

Warburgia breyeri > *Warburgia salutaris*

Zornia tetraphylla > *Zornia tetraphylla* var. *capensis*

General index

Index of plant names